INSTITUTIONAL CHOICE IN IRRIGATION

About the Author

Dr. Neetha N. is a Senior Fellow at the Centre for Women's Development Studies (CWDS), New Delhi. She holds a Ph.D. (Economics) from Centre for Development Studies/Jawaharlal Nehru University. She is a recipient of *IWMI-Ford Foundation Fellowship for South Asian Women Doctoral Scholars.* Prior to her joining CWDS, Dr. Neetha was Associate Fellow and Coordinator, Centre for Gender and Labour at the V.V. Giri National Labour Institute, NOIDA. She has many years of research experience in the fields of irrigation institutions, labour and gender studies and has several research papers published in reputed journals and books.

Institutional Choice in Irrigation

A Case Study of Distribution in A Command Area in Kerala

Neetha N.

CONCEPT PUBLISHING COMPANY PVT. LTD.,
NEW DELHI-110 059

The publication was financially supported by the ICSSR. The responsibility for the facts stated, opinions expressed or conclusions reached is entirely that of the author and the ICSSR accepts no responsibility for them.

The work was evaluated by Professor R. Parthasarthy, Director, Gujarat Institute of Development Research, Ahmedabad, consultant, on behalf of the ICSSR,

ISBN-13 : 978-81-8069-669-5 **ISBN-10 : 81-8069-669-3**

First Published 2010

Published and Printed by

Concept
Since 1974

Concept Publishing Company Pvt. Ltd.
Regd. Office :
A/15-16, Commercial Block, Mohan Garden
New Delhi-110059 (India)
Phones : 25351460, 25351794, *Fax :* 091-11-25357109
Email : publishing@conceptpub.com,
Website : www.conceptpub.com

Editorial Office :
H-13, Bali Nagar, New Delhi-110 015, India.

Cataloging in Publication Data-- *Courtesy:* D.K. Agencies (P) Ltd. <docinfo@dkagencies.com>

Neetha, N., 1970-
Institutional choice in irrigation : a case study of distribution in a command area in Kerala/Neetha N.
p. cm.
Originally presented as the author's thesis (Ph. D.)--Jawaharlal Nehru University.
Includes bibliographical references (p.).
Includes index.
ISBN 13: 9788180696695 ISBN 10: 8180696693

1. Chalakudy River Diversion Scheme. 2. Irrigation projects--India--Kerala. 3. Irrigation--Government policy--India--Kerala. 4. Irrigation canals and flumes--India--Kerala. I. Title.

DDC 333.913095483 22

To the Memory
of
My Father

Acknowledgements

I owe deeply to all, who have directly or indirectly contributed to the completion of this book, which is based on my Ph.D. dissertation carried out at the Centre for Development Studies, Jawaharlal Nehru University.

My foremost debt is to my supervisors, Dr. K. Pushpangadan and Dr.V. Santhakumar, whose intellectual contributions and enthusiastic interactions were crucial for me to carry out this work. Dr. Santhakumar was kind enough to spare considerable time in every stage of the study, including the fieldwork. Academic discussions with faculty and friends at CDS have enriched my work. I gratefully acknowledge the comments received from Dr. K.P. Kannan, Dr. G. Omkarnath and Dr. K. Narayanan Nair.

I am grateful to *Reserve Bank of India* for the Doctoral Research Fellowship awarded to me at Centre for Development Studies. I also acknowledge the support from *IWMI, Ford Foundation Fellowship for South Asian Women Doctoral Scholars.* In fact, this fellowship was crucial in the completion of my studies, due to the academic exposure that I gained from the detailed interactions with scholars at IWMI.

I also thank my external evaluators Prof. Tushar Shah and Prof. K.V. Raju for their encouraging comments and suggestions for bringing out the work in the form of a book.

My sincere thanks to the Indian Council of Social Science Research (ICSSR) for the award of publication grant, which enabled the publication of my research in its present form. I am deeply indebted to the ICSSR Consultant, Prof. R. Parthasarthy, Director, Gujarat Institute of Development Research, Ahmedabad, who evaluated the thesis on behalf of ICSSR. Thanks are also due to Dr. Prem Prakash Pandey of

ICSSR for all the support. I also thank the Concept Publishing Company Pvt. Ltd., New Delhi—Mr. Ashok Kumar Mittal and his team—for printing and publishing this book.

My greatest indebtedness is to Babu, whose selfless support, care and academic companionship was crucial in the completion of this book. Our sons, Appunni and Kannunni, took off most of my tensions and worries through their innocent smiles. I owe a lot to them for coping with my moods and hectic schedules, without any complaints.

I could not have completed this book but for the deep understanding of my families at Ponnani and Changanassery. Mottu and Amma were always there to enquire the day-to-day ups and downs and to instil me with confidence. During the course of the research work for this book, I lost my father, without whose direction and 'unexpressed' encouragement, I would not have been into social science research. I dedicate this work to his memory.

Neetha N.

Contents

List of Tables

List of Figures

1
Introduction

The recent decades have seen a large number of changes in the development and management of irrigation systems, the world over. Institutions and policies, which have governed the development and management of the world's irrigation systems in the past, have been found to be increasingly non-functional. The emphasis on state ownership and management of irrigation is waning with the emergence of new institutions.

During the early 1950s, development of irrigation was one of the prime areas of concern in most of the developing countries. This was reflected in the massive investment on the infrastructure undertaken in these countries and the world in general. During this period and till mid 1970s, termed as the *Construction Era* (Svendsen and Meinzen, 1997), public provision (through state ownership, operation and management) was considered as the most appropriate institution in irrigation.

The mid 1970s saw a shift in emphasis from irrigation development to effective management. This was due to: (a) the realisation that the benefits generated by existing irrigation systems were much below the expected level; and (b) the rising cost of new irrigation systems, which forced the Governments to focus more on irrigation management issues than fresh investments. This led to increased investment in system rehabilitation and the introduction of measures to improve the performance. System administration, focusing on both bureaucracy and water users, was the thrust area of the period, categorised as the *Improvement Era*. Notably, while attempts were made to effect changes in the superficial structures of the system, the basic institution remained the same.

Towards the end of the 1980s, the attention has shifted from structure improvement to the development of alternative institutions. Underlying the dominant thinking during the period, termed broadly as *Reform Era*, was the assumption that unless basic rules and structures governing irrigation management are changed, there is no possibility of improvement in irrigation development and management. Irrigation management transfer gained wider recognition during this phase, which focused on the transfer of all major responsibilities (water acquisition, distribution, maintenance, resource mobilisation and conflict resolution), from the government to the users. The period was also marked by the emergence of informal water markets. Such market-oriented solutions and privatisation of functions previously performed by the state, became part of the discourses on irrigation development.

The Context

Irrigation development in India is also marked by the three distinct phases of *Construction, Improvement* and *Reform*. A central feature of India's agricultural planning right from the First Five Year Plan has been the development of irrigation potential in the country. This was evident in the massive investment towards the creation of large-scale canal irrigation networks. During the period, 1951-2000, out of the total allocation of Rs. 2,19,164 crore on irrigation and flood control, Rs. 132,390 crore (60.41 per cent) was towards the development of major and medium projects[1].

During the first phase of irrigation development, a number of large-scale gravity irrigation projects were initiated and large area was brought under irrigation aimed at achieving self-sufficiency in foodgrain production. Public investment in large-scale projects was viewed desirable on account of issues related to natural monopoly, massive investment requirement and associated economies of scale. Till 1970s, irrigation was managed and administered entirely by the state. Distribution was also vested entirely with the state and co-operation of farmers was enforced through sanctions and penalties.

However, the initial thrust on bureaucratically managed irrigation systems received a setback due to the failure of the institution to meet the envisaged targets and objectives. A number of studies reveal administrative flaws leading to mismanagement and rent seeking and nepotism by irrigation officials and politicians, in the state owned and managed irrigation systems (Wade, 1975, 1985, 1988a; Chambers, 1988, Repetto, 1986). Consequently, there has been a growing dissatisfaction with the centralised system of irrigation management (Wade 1978, 1980; Chambers, 1988, World Bank, 1994).

An analysis of the shortcomings of the conventional irrigation management points to the lack of involvement of farmers in irrigation management. By early 1970s, it was generally felt that the major problems of canal irrigation management are located 'below the outlet'. Institutional problems were largely conceived as outcomes of lack of participation by the farmers in the construction and maintenance of watercourses. This necessitated the introduction of measures to encourage farmer participation in irrigation management as complementary to public provision. Consequently, Command Area Development (CAD) Programme was initiated in 1973, focusing on user-based associations in tackling irrigation problems 'below the outlet'. Under this Programme, farmers contributed to the construction and maintenance of field channels and were associated with water management below the outlet. However, by early 1980s' it was found that the rotational water supply system and water users associations introduced under the CAD Programme failed, after the initial impetus (Singh *et al.*, 1994).

During 1980s, there was a growing understanding of the fact that apart from the institutional problems 'below the outlet', there exist problems 'above the outlet' also. Thus, considerable efforts were made to involve users in the management of surface irrigation systems both above and below the outlet. This led to the development of a new concept of participatory irrigation management, whereby user groups were involved in the management of irrigation systems, with a limited role of the state. Accordingly, one of the objectives of Water Management Policy of the Government of India (Government of India, 1987) was the phased transfer of the tertiary system management to water users' associations. Although much has been envisaged about management transfer in irrigation to user association, actual changes in the field have not been extensive. Also, such transfer attempts have varied greatly, across states and projects (Brewer *et al.*, 1999a, 1999b).

Apart from the state-led institutional changes, a number of alternatives have also emerged, as farmer's responses to the inefficiencies in public provision. Many non-governmental initiatives of collective action have been reported in canal commands, which are either complementary to or substitute for state managed systems. These collective actions take varied forms such as fact finding, lobbying, raiding, guarding, operating controls, construction of watercourses, water capture and maintenance, and design and implementation of turns (Chambers, 1988). Construction and maintenance of community wells, ponds and other water reservoirs are the other areas where farmers' cooperative initiatives (without much state support) have been documented. Apart from these collective actions, the development of ground-water markets has been pervasive[2] in canal commands. Due to problems in canal irrigation, sale of water from wells are found spreading in the command areas of surface irrigation projects (Shah, 1993, 2001). The development of water markets is mainly associated with the spread of tubewells or pump irrigation. There are also many cases of sale of irrigation-turns by the farmers. Increased private investment for irrigation by the farmers in the form of private wells and pumps, eventually with an intention to counter uncertainty about and inadequacy in water availability under public provision are also reported extensively. Spread of government aided or community based lift irrigation schemes is also noticed in the command areas of the canal irrigation schemes (Jacob, 1996). Water is lifted or diverted from rivers, ponds, or reservoirs for irrigating water-scarce areas in the canal command, under the aegis of farmer collectives, community organisations or government.

All these institutional developments explained above, indicate the existence of multiple institutions in the provision of irrigation in the canal command, which have developed as the result of continuous interactions between various actors within the existing institutional framework of public provision. It is important to understand the evolution and functioning of these institutions, as it gives important insights into the dynamics of institutional change and its determinants. An analysis of farmers' choice of institutions from multiple alternatives could also help in revealing the constraints affecting the choice of various actors. The process of evolution and the efficiency and equity aspects of these institutions are also important in framing policies on irrigation management.

In this context, the book focuses on the emergence and working of various institutions involved in the provision of irrigation, and analyses the farmers' choice of institutions in the command area of a major surface irrigation project—Chalakkudy River Diversion Scheme (CRDS) in Kerala. It analyses the development and functioning of various institutions in the canal command that have emerged as outcomes of farmer responses to the failure of public provision.

Approach of the Study

The most important justification for state intervention in the management of irrigation is provided by the argument that without state intervention (or without coercion) people would not cooperate in realising their common interests. The failure of farmers to voluntarily share the cost of provision and management of irrigation, which has some features of public good, and the economies of scale (or natural monopoly) associated with the production and distribution of irrigation, reinforce the case for state interference. However, the past experience shows that there can be state-mismanagement of irrigation leading to failures. The state mediated production and distribution need not be the most beneficial form of public intervention and that non-state collective actions are possible to overcome the problems of public good and diseconomies of scale.

Interactions between individuals in the context of irrigation are dynamic. The choice of an individual farmer differs from that of others and can vary over time. However, at a given point of time, there may be some clear-cut patterns of choice, which may be governed by formal or informal institutions. Studies on irrigation institutions have defined the term, *institution* as a set of rules in use to organise repetitive actions by individuals that produce outcomes, which affects those individuals and potentially others (Ostrom, 1990). All irrigation institutions are crafted with considerations of effectiveness, efficiency, equity and sustainability in the provision of water. These institutions are largely shaped by the prevailing social and economic environment.

In the context of the present study, each set of formal or informal rules or norms governing the modes of acquiring irrigation water such as collective action with and without government support, water market, private investment in open wells in individual farms and so on, is considered as an institution. Institutional choices in irrigation can be analysed from the perspective of an individual farmer who is assumed to be maximizing pay-offs. The individual's choices of institutions in any situation will be a function of expected benefits and expected costs, given other constraints. An individual farmer evaluates benefits and costs from an alternative set of rules *vis-à-vis* the *status quo*. The individuals cost-benefit calculations, as regards institutional change broadly depend on : (a) the benefit from the increased amount of irrigation water provided as a result of shifting to a new institution; (b) the fixed and recurring cost involved in the new institutional framework (both monetary and non-monetary).

For plots with given crop in a locality, one can assume that there will not be much difference in benefits from irrigation if the technology and the use of other inputs in crop production are by and large similar. Thus, the choice of irrigation institution depends on the cost. Costs (*i.e.*, over and above the Government fixed cess) associated with various institutions vary across individuals and plots. These costs can be broadly divided into two : production and transaction costs.

Transaction costs are the costs of measuring the valuable attributes of what is being exchanged and the costs of protecting rights and policing and enforcing agreements (North, 1990). With positive transaction costs associated across various institutional alternatives of acquiring irrigation water, economising takes place not with reference to production costs alone but with total costs, which is the sum of production and transaction costs. In other words, a farmer selects a particular institution that minimises the total costs (production costs + transaction costs) thereby maximising the net payoffs.

Here the institutional choice at the distributary level reflects the economic calculations of individual farmers. However, because of the potential economies of scale in water distribution, decentralised decisions and investment need not always be efficient in aggregate terms. Thus, it is important to analyse the impact of the aggregate of these decentralized decisions on aggregate efficiency and the distribution of income.

As mentioned earlier, the study was carried out in the command area of the Chalakkudy River Diversion Scheme (CRDS), which is spread over the districts of Thrissur and Ernakulam in Kerala. The CRDS is selected, because of the available evidence that multiple institutions exist in the canal command. Further, it is also the first major surface irrigation project in the state (commenced in 1957), which provides ample scope for an *ex post* study. The study analyses the mode of acquiring water in 397 plots owned by 210 households. The sample is selected giving adequate representation to various reaches of the canal.

Chapter Scheme of the Book

The book is organised in 8 chapters. Chapter I introduces the book and explains the context, scope and approach of the study. Chapter II provides an overview of irrigation development in India and Kerala and examines the failure of public provision, which forms the background of the study. Chapter III reviews the theoretical approaches to the study of institutions. It also examines the empirical literature on irrigation institutions and identifies the constraints affecting irrigation decisions of farmers and discusses the framework for analysing such institutions in the study. A brief profile of the CRDS and the selection of the sample is explained in chapter IV. A detailed examination of various institutions existing in the canal command of CRDS is also given in this Chapter along with case studies. Chapter V analyses various constraints in the institutional choice of farmers and delineates the central variables. The centrality of cost considerations in institutional choice and the various components of cost across various irrigation institutions are analysed in Chapter VI. The efficiency and equity issues of public provision and alternative institutions are discussed in Chapter VII. Finally, Chapter VIII summarises the findings and suggests policy prescriptions.

NOTES

1. *Indian Planning Experience : A Statistical Profile,* Planning Commission, Government of India (2002). The figures indicate the expenditure at 1996-97 prices.
2. The issue received wider attention since the pioneering works of Shah (Shah, 1985, 1989, 1993).

2

Background of the Study

Canal Irrigation in India and Kerala

Recognising the vast potential of irrigation in enhancing agricultural production, considerable investments have been made by less developed countries in creating irrigation infrastructure, especially large surface projects. Notwithstanding these massive investments, low irrigation efficiency, low productivity levels and inequitable distribution of water characterise canal irrigation in most of the developing countries. This dismal situation is often attributed to inappropriate design and management, improper financial planning, inadequate maintenance, lack of farmer participation, undesirable political intervention and so on. Though these issues are common to all irrigation projects, the gravity of the problem varies across countries and projects. This necessitates discussions of issues in specific regional contexts.

Canal Irrigation in India : An Overview

Massive investment in irrigation infrastructure (in both surface and ground water systems) has been a major development priority and strategy in India. Ever since the Independence, the State and Central Governments have made large investments in construction of government operated irrigation systems (Brewer *et al.*, 1999a, Pant, 1981). The primary justification given for this is that large irrigation systems demand greater and more centralised control and government often has higher resources combined with the requisite technical knowledge and expertise for managing them. Accordingly, larger surface irrigation systems attracted greater attention and financial allocation from Central and State Governments.

The priority accorded to irrigation in the post-Independence era resulted in a visible increase in the irrigation potential. Within the first four decades of planned development, a considerable share of government's outlay was channelled to build up irrigation infrastructure, resulting in a four-fold increase in irrigated area and the country, now, has the second largest irrigated area in the world (Palanisami, 1984). About 65 per cent of the cumulative outlay on irrigation during 1951-97 was spent on large and medium projects, which command about 40 per cent of the total irrigation potential of the country (Table 2.1).

Table 2.1 : Investment in Irrigation Sector (1951-1997)

(Rs. crore)

Period	*Total Plan Expenditure*	*Investment in Irrigation*	*Percentage of Irrigation Investment to Total Plan Expenditure*	*Invest-ment on Major and Medium Projects*	*Percentage of Invest-ment on Major and Medium Projects to Total Irrigation Investment*
1951-56	1360	446	32.79	380	85.20
1956-61	4672	541	11.58	380	70.24
1961-66	8577	1024	11.94	581	56.74
1966-69	6603	995	15.07	434	43.62
1969-74	15778	2411	15.28	1237	51.31
1974-78	28811	3853	13.37	2442	63.38
1978-80	22941	3043	13.26	2056	67.56
1980-85	109646	10948	9.98	7531	68.79
1985-90	180000	17874	9.93	11556	64.65
1990-92	125835	8949	7.11	5347	59.75
1992-97	434100	33510	7.72	22214	66.29
Total	938323	83594	8.91	54158	64.79

Source : Central Ground Water Board, New Delhi.

Distribution of surface water irrigation is perceived primarily as a form of welfare, and thus irrigation systems are designed to spread water to as many farmers as possible. This is also consistent with one of the goals of irrigation management in all States, which is to ensure equity in the distribution of water to all recognised users. Public irrigation systems (through canal), deliver water to outlets, each of which serves more than one farmer. Below each outlet, farmers are collectively responsible for both water distribution and maintenance of the distribution system. Outlets generally serve areas ranging from 5 to 300 hectares and from 5 to 100 farmers. By and large, four main approaches have been followed for water distribution in India—*Warabandi*[1] (in Northwestern India), *Shejpali*[2] (in Western India), *Land Classes Approach*[3] (in Southern India) and *Satta System/Assured Irrigation*[4] (in Eastern India).

The growing problems of low water use efficiency and financial sustainability of the systems raised serious doubts about the efficacy of the public provision and management of irrigation. By 1970s, it was increasingly recognised that the "irrigation potential created" for most government-managed irrigation systems was far in excess of the "potential realised". This clearly meant that most of the government-managed irrigation systems were unable to irrigate their designed commands (Pant, 1981).

Numerous studies have shown that these systems have performed poorly in terms of irrigating the planned commands, system maintenance, water distribution, equity and efficiency in water usage and recovery of water charges. Also, increased pressure on government finances has resulted in the further deterioration of the irrigation systems, as less and fewer resources are devoted to system maintenance (Wade, 1980; Pant, 1981; Palanisami, 1984 ; Mathur and Saibal 2000; Reddy, 1997). While commenting on the generally disappointing performance of large publically operated canal systems and the sizeable gap between the irrigation potential "created" and "realised", Wade (1980) observes that canal systems are designed and constructed assuming a level of utilisation of control capacity which does not exist in most systems in South Asia.

The failure of the large scale irrigation systems sparked off debates on the comparative efficiency of major systems against minor irrigation projects. Based on an analysis of plan outlays for various types of irrigation and their respective performances, Reddy (1997) establishes that minor irrigation sources have been more efficiently utilised than major and medium irrigation sources. While major and medium sources, accounting for more than 5 per cent of cumulative plan outlay, created only 37.9 million hectares of irrigation potential, for minor irrigation, it was 61 million hectares (with only 1.38 per cent of plan outlay).

A widely cited and important reason for the unimpressive performance of canal irrigation is its poor maintenance of the system due to inadequate resources. The poor recovery of costs through the water cess levied on users, in most cases, could not even cover the Operation and Maintenance (O&M) charges of the systems, let alone the construction costs. With the progressive expansion of government-managed irrigation following independence, the need for resources for Operation and Maintenance (O&M) of the systems grew rapidly. Nevertheless, the government, which viewed irrigation purely as a welfare measure, has been reluctant to raise irrigation fees charged to farmers. In addition, as irrigation services declined and irrigation agencies weakened, farmers became reluctant to pay even low irrigation fees. This reduced the resources available for financing irrigation development leading to inadequate maintenance and system deterioration, as early as late 1960s. In early 1970s, the Second Irrigation Commission strongly recommended raising irrigation fees to cover O&M cost. However, the states took little action, probably because such a step was politically unpalatable (Brewer *et al.*, 1999a).

The Second Irrigation Commission recognised the fact that bringing the 'potential realised' closer to `potential created' required exploring ways to spread water to greater portions of the designed command areas, while ensuring the availability of sufficient water for the crops. This provided a stimulus to efforts at evolving measures for improved water use efficiency. Farmer participation in management and distribution was the perceived solution to the problem below the outlet. Accordingly, Command Area Development (CAD) Programme was initiated since early 1970s, by the State governments to promote farmer participation below the outlet. This was followed by programmes aimed at encouraging farmer participation in system management above the outlet—an idea that got increased recognition by early 1990s (Pant, 1981; Parthasarathy, 2000). However, these programmes were not successful in substantially improving the performance of the projects (Joshi and Hooja, 2000; Singh, 2000).

By 1990s, the problem of financing irrigation, O&M had grown sufficiently acute prompting the Planning Commission to constitute a committee on pricing of irrigation water. The major recommendation of the Commission was to increase water fees so that the O&M are met. The committee also pointed out that the O&M finance problems were linked to system performance problems and to the management capabilities of irrigation agencies. The report advocates "a substantial reduction in the sphere of responsibility of government" and "the encouragement of user groups to take over maintenance, management of water allocations and collection of water rates". The above discussion makes it clear that canal irrigation systems in India are under considerable strain because of the deteriorating condition of the physical system, poor cost recovery and lack of farmers involvement in the management of irrigation.

Surface Irrigation in Kerala : A Case of Failure

The irrigation requirement of Kerala is distinct from the rest of India. The State is blessed with relatively timely, adequate rainfall with an annual rate of 3107 mm. The South-West Monsoon (June—September) contributes 66 per cent, the North-East (October-December) 16 per cent, the Winter rains (January—February) 3 per cent, and Summer rains (March—May) 15 per cent[5]. The State does not suffer from too wide an inter-annual variation in the rainfall. However, large deviations occur in monthly rainfall and rainfall across regions, which makes irrigation a necessity for stabilisation of the water requirement of the crops. Irrigation in the State mostly has the status of protective irrigation, the focus being on the efficient management of water.

Irrigation becomes all the more important in the context of Kerala since 90 per cent of the State's cultivable area has already been brought under cultivation and the only way to increase production is intensive cropping through increased use of inputs such as irrigation. Accordingly, various Five-Year Plans of the State made large allocations for irrigation (Table 2.2). Irrigation is the third largest sector of investment in the State, with a cumulative investment amounting to Rs. 2,735 crore, till the end of 1999-2000. Out of this, 68 per cent (Rs. 1,877 crore) was on major and medium projects. The total irrigation potential created so far is 3.80 lakh ha. (net) and 4.70 lakh ha. (gross)[6].

Table 2.2 : Plan-wise and Source-wise Investment in Irrigation in Kerala

Plan Period	*Expenditure (Rs. Crore)*			*Percentage Share of Major and Medium Projects*
	Major/ Medium Irrigation	*Minor Irrigation*	*Total Irrigation*	
First Plan (1951-56)	11.79	—	11.79	100
Second Plan (1956-61)	7.91	2.07	9.98	79.26
Third Plan (1961-66)	10.29	7.20	17.49	58.83
Annual Plans (1966-69)	9.16	7.87	17.03	53.79
Fourth Plan (1969-74)	27.36	21.18	48.54	56.37
Fifth Plan (1974-78)	75.13	31.15	106.28	70.69
Annual Plans (1978-80)	74.97	29.09	104.06	72.04
Sixth Plan (1980-85)	259.53	58.94	318.47	81.49
Seventh Plan (1985-90)	301.90	137.71	439.61	68.67
Eighth Plan (1990-95)	375.00	130.00	505.00	74.26
Ninth Plan (1997-2002)	650.00	250.00	900.00	72.22

Source : *Plan Documents*, Government of Kerala.

Kerala is not an exception to the 'big dam culture' in India. The irrigation investment in the State mirrors the national pattern of massive investment incurred on major and medium irrigation projects, as a part of the agriculture development policy in the Post-independence period. The underlying assumption behind promoting large scale irrigation projects was that the extensive enhancement of command areas could alter the crop pattern as well as the intensity of cropping and realise higher levels in the productivity and production of food crops.

Despite the massive investment in irrigation, the achievement of the major and medium projects has been far from expected as is the case in the rest of the country. The marked gap between irrigation potential created and utilised and the sub-optimal use of water resources are indicative of the poor performance of the State in the field of irrigation management. Even in the completed irrigation projects, the full benefits have not been realised (Table 2.3).

In the case of the completed projects, the overall physical achievement realised from the major and medium projects is only 73 per cent of the expected potential. As regards the ongoing projects, only 20 per cent of the targeted area has been brought under irrigation, despite heavy investments incurred on them. Improper financial planning, cost escalation, inordinate delay in construction are among the key factors that accounts for the poor physical performance of the projects (Netto, 1990; Viswanathan, 1997; Santhakumar, 1997).

The impact of irrigation on farm output has been one of the key areas of interest. Irrigation is undoubtedly a critical input in production and yet its contribution in the performance of the agricultural sector of the State has been rather unsteady. A number of studies have revealed that there is no clear evidence to show that irrigation projects have significantly benefited Kerala's agriculture and that farm output has increased substantially with the provision of irrigation. In fact, many have pointed out that, given the current trend and pattern of agricultural development, the scope of irrigated agriculture is waning (George and Nair, 1982; Narayana and Nair 1983; Kannan and Pushpangadan, 1989).

Table 2.3 : Physical Performance of Major and Medium Irrigation Projects in Kerala

Name of the Project	*Physical Target (Area in ha.)*	*Physical Achievement (Area in ha.)*	*Achievement (%)*
	A : Completed Projects		
Chalakkudy	19690	13530	68.71
Peechi	17555	15262	86.94
Malampuzha	29463	19802	67.21
Neyyar	16042	8300	51.74
Pothundy	8792	4685	53.29
Gayathri	7651	4880	63.78
Walayar	4536	3752	82.72
Vazhani	3565	2113	59.27
Mangalam	4816	3313	68.79
Cheerakuzhi	2268	952	41.98
Pamba	21135	20710	97.99
Periyar Valley	32800	30567	93.19
Chitturpuzha	15700	16102	102.56
Kuttiadi	14570	1411	9.68
Grand Total	198583	145379	73.21
	B. Ongoing Projects		
Kallada	61630	35602	57.77
Chimony Dam	13000	1300	10.00
Kanjirapuzha	9713	7266	74.81
Pazhassi	11525	6348	55.08
Muvattupuzha	17737	—	—
Vamanapuram	8800	—	—
Idamalayar	14394	—	—
Kuriarkutti	17488	—	—
Chaliyar	73240	—	—
Kakkadavu	13940	—	—
Attapady	4347	—	—
Karapuzha	5521	—	—
Meenachil	9960	—	—
Banasurasagar	2800	—	—
Chamravattom	3106	—	—
Kanakkankadavu	2600	2600	100
Trithala	1303	—	—
Grand Total	260280	53116	20.41

Source : *Economic Review* (various years), State Planning Board, Government of Kerala.

Against this backdrop, the subsequent discussion reviews the failure of the canal system in the State, with a focus on the bureaucratic mis-management and flaws in irrigation policy of the State. It is argued that irrigation development in the State has largely neglected the specifics of Kerala's agriculture influenced and determined by a host of factors such as the agro-climatic and agro-ecological patterns, changes in size of holdings, and the overall dynamics of the socio-economic and political setting. The over-emphasis on paddy cultivation, which is still the dominant feature of the agricultural policy of the State, is critically examined in the context of the rapid change in cropping pattern towards perennial and cash crops.

Neglect of Agro-climatic Factors in the Design

A major criticism of major irrigation projects in Kerala has been that these projects are looked upon as mere engineering tasks while neglecting the agro-ecological features of humid-tropic Kerala and the specific characteristics of its agrarian communities (George and Nair, 1982; Santhakumar, 1997). The contention that Canal Irrigation Systems, based on a strong technological edifice, are the only means for irrigation in a high rainfall area like Kerala, has been questioned by Santhakumar *et al.* (1995). It is held that the bias towards major projects is the result of a coalition between the political leadership and technocratic systems. This is evident from the fact that a realistic estimate of the availability of water locally is not considered in the planning and designing of projects. Further, it is maintained that, in their endeavour to achieve the political objective of self-sufficiency in rice production, the policy-planners too opted for dam and canal based systems, ignoring the fact that these are 'naturally' costly, in a State like Kerala, given the possibilities of meeting the relatively small requirement of additional water through less expensive means.

Excessive Importance to Paddy Cultivation

A major flaw in agricultural and irrigation policies of the State is their exclusive orientation to paddy cultivation. Kerala's agriculture stands apart from the rest of the country. A considerable proportion of crops are grown as garden crops in small plots of land. The agro-climatic situation permits viable cultivation of mixed and multiple crops in the State, due to which, the cropping pattern consists of a mix of crops though regional differences exist in the mix. While referring to the natural advantage of mixed cropping in the region, Nair (1999) observes that the undulating topography as well as moisture availability on different facets at any given point of time varies even within small farmsteads enable the farmers to grow multiple crops.

Despite this natural advantage, however, the State has invested massively on surface irrigation projects aiming at enhanced production of paddy. This was done in pursuance of the national political agenda of "Grow More Food Campaign" aiming at food self-sufficiency—a policy response to the Second World War food shortages. In addition to the food security aspect of canal investment, paddy promotion was also thought to generate higher employment for agricultural labourers (Santhakumar and Nair, 1999). Accordingly, considerable emphasis was put on irrigation investments for paddy production under the Five-Year Plans in the State (Santhakumar *et al.*, 1995). Crops other than paddy were considered for irrigation only during 1980s.

The introduction of an irrigation technology that goes against the natural selection of crop mix, had undesirable implications for the food scenario in Kerala. Contrary to the expectation that the promotion of 'paddy-oriented' canal irrigation would improve the food situation, the fall in acreage under paddy during the period was alarming, questioning the very logic of irrigation planning. The production of foodgrains (primarily rice) declined from 1.3 million in early 1980s to 1.1 million tonnes in mid 1990s to 0.7 million in 2002-03 bringing down the share of State in food production in the country from 1.01 to 0.70 per cent of the country. The ratio of domestic production of rice to total rice requirement in the State has declined from 0.45 to 0.12 during this period.

Significantly, large scale irrigation projects in Kerala were not successful in enhancing farm output, by bringing additional area under cultivation or increasing cropping intensity (Pillai, 1982; Narayana and Nair, 1983; Kannan and Pushpangadan, 1988). Further, is noted that irrigation in Kerala has not led to any increase in productivity but has only helped in stabilising production through supplementing the existing source of water (George and Nair, 1982). Kannan and Pushpangadan (1989), however, argue that even in Palakkad district, where public investment on irrigation has been the maximum, there has been no evidence to show that irrigation has either stabilised or increased crop intensity. The poor performance of irrigation has been attributed to the pattern of rainfall and topography, which imposed severe restrictions on the absorption of modern technology in rice (Santhakumar *et al.*, 1995).

The excessive emphasis placed on paddy promotion in the irrigation policy was largely detrimental to the development of other crops. Nair (1999) observes that this wrong move resulted in the sinking of massive resources in low return investments (such as irrigation) and the sub-optimal use of scarce land and water resources, which together discouraged or prevented the development of other crops. Further, various measures adopted by the Government including the Land Utilisation Order, 1967[7], under the Essential Commodities Act, 1955, discouraged conversion of paddy and prevented the allocation of the scarce land and water resources to more income generating and employment generating activities such as production of vegetables, bananas and so on. In effect, the undue stress on promoting mono-crop farming (of paddy) is proved to be counter-productive.

Imposition of a Static Frame on a Dynamic Milieu

The design of surface irrigation projects was static, juxtaposed on a dynamic milieu of changing agrarian relations and socio-economic environment of the State. In other words, the changes in the cropping pattern, size of holdings, occupational pattern, availability of alternative sources of irrigation and relative profitability of crops were not considered in the planning of irrigation projects.

The cropping pattern of the State had undergone phenomenal shifts since the mid 1970s, favouring non-food-crops. The factors that contributed to this shift include relative price movements of crops, advent of irrigation, increasing cost of cultivation, fragmentation of land and so on. Though irrigation is also a factor that contributed to the expansion of non-food crops, in the context of the study, this expansion is considered as a departure from the objectives of the canal based surface irrigation projects, which are essentially designed for augmenting paddy production in the state.[8] As a result, rice production, which was at its peak of around 14 lakh tonnes in the mid 1970s, declined drastically to 6.9 lakh tonnes in 2002-03. Further, consequent to the enormous pressures from high value crops like coconut, banana, pineapple and plantain the area under paddy declined from its peak coverage of 8.81 lakh ha. in mid-seventies to 3.1 lakh ha. in 2002-2003 (Table 2.4). Two major trends are evident. Firstly, there is a visible shift from paddy to other crops. The area under paddy has consistently declined since 1980s and the rate of decline was higher in the 1990s. Secondly, the crops, which have expanded considerably, are banana and plantain, coconut, rubber and arecanut. Of these, the only crop for which the area under cultivation has consistently grown is banana and plantain. It may be noted that there is decline in the area under coconut and rubber since mid-1990s.

The shift in cropping pattern, in conventional terms, can be explained in terms of movements in relative prices of various crops, given the provision of adequate irrigation water. The reduction in the area under paddy, thus, can be explained in terms of increases profitability of alternative crops (coconut, plantain, and rubber). In the case of shift from paddy to coconut, there has been a sharp move from seasonal to perennial crops, with long term implications. The shifts in this regard will not only be a function of current relative prices but also of the future expected prices and additional investment required. However, plantain is a seasonal crop and (unlike coconut) can be an immediate substitute with low investment. It is seen that while the prices of paddy increased only 178.49 per cent during the period 1970-71 to 1987-88, the increase for banana, coconut, rubber, pepper, tapioca was more than 300 per cent (Jose, 1991), which explains the farmer's preference for non-paddy crops.

Table 2.4 : Trends in Crop-wise area in Kerala, 1960-61—1999-2000

(Area in '000 ha)

Crops	*1960-61*	*1970-71*	*1980-81*	*1990-91*	*1995-96*	*1999-2000*	*2002-2003*
Paddy	778.91	874.93 (1.17)	801.7 (-0.87)	559.45 (-3.53)	471.15 (-3.38)	349.77 (-6.44)	310.52
Coconut	500.76	719.14 (3.69)	651.37 (-0.98)	864.06 (2.87)	982.11 (2.59)	925.04 (-1.45)	905.48
Tapioca	242.2	293.55 (1.94)	244.99 (-1.79)	147.25 (-4.96)	118.74 (-4.21)	111.92 (-1.44)	110.3
Banana and Plantain	44.42	48.76 (0.94)	49.26 (0.1)	61.16 (2.19)	73.8 (3.83)	92.30 (6.27)	107.12
Rubber	122.87	179.26 (3.85)	237.8 (2.87)	384 (4.91)	523.24 (6.38)	472.90 (-2.41)	476.05
Arecanut	54.26	85.82 (4.69)	61.24 (-3.32)	62.06 (0.13)	76.53 (4.28)	81.94 (1.77)	92.59
Pepper	99.75	117.54 (1.65)	108.07 (-0.84)	168.99 (4.57)	190.83 (2.46)	198.41 (0.99)	201.04
Cashew	54.32	102.71 (6.58)	141.27 (3.24)	118.09 (-1.78)	114.82 (-0.56)	89.40 (-5.53)	86.62
Coffee	16.8	31.56 (6.51)	57.56 (6.19)	84.01 (3.85)	82.34 (-0.4)	84.14 (0.55)	84.14
Tea	37.61	37.59 (-0.01)	36.16 (-0.39)	34.61 (-0.44)	35.03 (0.24)	34.79 (-0.17)	36.82
Pulses	44.12	39.94 (-0.99)	33.86 (-1.64)	24.37 (-3.24)	17.12 (-6.82)	10.99 (-8.95)	7.36
Ginger	12	12.17 (0.14)	12.66 (0.4)	14.04 (1.04)	11.2 (-4.42)	11.26 (0.13)	10.37

Source : *Economic Review* (various issues), Government of Kerala.
Note : Figures in brackets are the annual compound growth rates.

Yet another factor considered important in explaining the shift away from paddy was the increasing cost of cultivation

(Unni, 1983; George 1982; Panickar 1980; Kannan and Pushpangadan, 1988). It was found that the cost of inputs for paddy cultivation, particularly the wages had risen disproportionately to the price of paddy. Apart from the rising labour cost, the acute shortage of labour[9] to carry out field operations, especially during peak seasons also forced the farmers to shy away from paddy to labour saving non-food crops (Francis, 1990). The shift in cropping pattern could also be seen as a reflection of the changing occupational status. Cultivation is no longer the primary occupation[10] of the landed households, but as a secondary occupation, and hence there exists a widespread preference for perennial and cash crops, which does not demand much time and labour compared, to paddy.

The sub-division of land into small holdings was also crucial in the State's agrarian transformation. Land reforms, decline of joint family system and the increased population pressure have resulted in proliferation of small-holdings. The following Table (Table 2.5) shows the distribution of land in the State across various size holdings.

Table 2.5 : Number of Operation Holdings and Area operated by Size Class in Kerala—1995-96

Size of Holding (in ha)	*Number of Holdings*	*Proportion to Total No. of Holdings*	*Area (in ha)*	*Proportion to Total Area*	*Average Size (in ha)*
Below 0.02	751951	11.94	10150	0.59	0.01
0.02-0.50	4683476	74.37	569248	33.19	0.12
0.50-1	483648	7.68	336425	19.61	0.70
Marginal (upto 1 ha)	**5919075**	**93.99**	**915823**	**53.39**	**0.16**
1-2	261418	4.15	346100	20.18	1.32
Above 2	116794	1.85	453381	26.43	3.88
Total	**6297287**	**100.00**	**1715304**	**100.00**	**0.27**

Source : *Economic Survey 2001*, State Planning Board, Government of Kerala.

The Table 2.5 reveals that about 94 per cent of the land-holdings falls in the category of marginal farms with an average size of 0.16 hectares. The proliferations of small holdings were found unable to sustain paddy cultivation as the crop is more suitable for large scale farming (Jose, 1991). Further, it is unrealistic to assume that such holdings could make necessary investment for land improvement or irrigation (Santhakumar and Nair, 1999).

These developments in the agrarian scene indicate that the scope of paddy farming has declined drastically defeating the very logic of irrigation planning in the State. As explained above the entire planning of surface irrigation projects in the State was narrowly focused towards the promotion of paddy cultivation. Thus, it failed miserably in foreseeing the transformations in the farm economy, especially the drastic shift in cropping pattern. This failure can be attributed to the shortsightedness and rigidities in designing irrigation projects.

Another important anxiety in the context of rapidly changing cropping pattern is the issue as to whether the existing systems of irrigation could support the requirements of non-paddy crops. Rosegrant *et al.* (1995) point out that the existing rice-based irrigation systems constrain diversification of crops because of the rigid design in infrastructure and inflexible water delivery systems and the difficulty and high cost of converting and using existing water distribution systems. Rice-based irrigation systems are primarily designed to supply water for wet season when, except for occasional short dry spells, irrigation management needs are less stringent unlike other crops.

The rigid frame of rice-based irrigation system failed to respond to the varying water needs of dry season cropping patterns due to many counts. Water control for non-rice crops is more demanding, though the total volume of water required for the former is less than the latter. Non-rice crops require intermittent or rotational water supply, whereas for rice it can be continuous or intermittent. Thus, a high degree of water control is often necessary to provide a reliable and timely intermittent water supply to non-rice crops (depending on the growth stages and types of crops being grown).

As has been seen, in the context of Kerala, the crops that have replaced paddy, are coconut, arecanut, banana, pepper and rubber. Of these, coconut and arecanut are perennial crops and their water requirement is limited to a few months in summer. Thus, the irrigation requirement of these crops is more for productive than protective purposes. Banana and pepper cultivation, though not so water intensive requires water on a much regular basis, which is also different from that of paddy. The critical question, thus, is will the existing institutional arrangement under public provision of canal irrigation be suitable to meet the challenges posed by the diverse water requirements of the various crops.

The delay in the planning and implementation of irrigation projects, an issue which is often cited in the discussions on the failure of major irrigation projects not only results in huge cost over runs but also leads to a mismatch between the requirements of cultivation and the envisaged objectives of the project. The shift in cropping pattern during the implementation period of the projects, which is often too long is rarely taken into account. Thus, irrigation projects designed for the requirements of paddy cultivation may have restricted use even in the initial periods of functioning of the projects, because of the changes in cropping pattern during the period of project implementation[11].

Administrative Flaws : Improper Management and Cost Recovery

Most of the canal systems in India are merely administered rather than managed and Kerala stands no exception to this. The State carries the responsibility of managing the large irrigation systems and the irrigation department is responsible for delivering adequate water till the crop is ready for harvesting. Inadequate and uncertain water availability has been a problem common to all the projects in the State. Water rights are assigned to land, with some land entitled for irrigating two crops, while the rest are permitted for single crop irrigation. The release of water to the outlet is not based on any systematic calculations, but is largely dependent on the availability of water, the total area under irrigation and the number of crops cultivated. The commonly cited issues of canal irrigation such as corruption and rent seeking among officials, mismanagement, lack of commitment and coordination, politically directed and personally motivated interests, frequent transfers of engineers and staff are cited widely in Kerala, though there are no concrete studies to substantiate these concerns.

In view with the general perception that lack of farmer participation deters proper and efficient management of irrigation in Kerala, attempts have been made in the late 1970s to promote community participation in managing canals. Accordingly, Irrigation Advisory Committees[12] and Sub-Canal Committees[13] were formed in selected projects which was subsequently replaced by Command Area Development (CAD) Programme[14] in 1980s. Nevertheless, the achievements of the Programme were considerably below targets. The administrative outlook of officials was again cited as one of the main reasons for the failure of participatory irrigation management in the State (Brewer *et al.*, 1999a). The absence of any commitment to participatory irrigation management among the State officials has evoked much criticism. Even the adoption of the policy for CADA systems was, it is felt, primarily a means of getting Central Government funds and to counter criticisms about the poor performance of investments in irrigation (Chackacherry, 1993).

The gross under-pricing of water is another issue, which is relevant in the context of improper functioning of the irrigation projects. The only financial return that accrues to the State from irrigation projects is the irrigation cess or water fees, which are collected as part of the land tax and are based on water rights given to land. The water rates have not been revised for about three decades, the latest being in 1974[15]. The highly subsidised irrigation charges and lack of concerted effort to collect it make irrigation almost free of cost. Further, there is no coordination between the Irrigation Department and the Revenue Department, which is entrusted with the task of collecting the water cess. Brewer *et al.* (1999a) finds that collection of irrigation fees through the land tax system is probably a detriment to effective irrigation management, particularly when the State cannot provide the funds needed for regular allocations.

To sum up, irrigation planning and management in the State, with its thrust on paddy-based agricultural development has failed to cope up with the changes in agrarian structure. It is evident that the basic assumption behind the construction of major irrigation projects, that all lands are fully convertible to paddy though provision of surface irrigation was proved to be faulty. The demand for irrigation has changed considerably, as the State has almost reached a saturation point in respect of land use for agriculture and the emerging trend is towards a shift in the cropping pattern in favour of less water demanding perennial crops in lieu of seasonal crops. All these changes in the farm front demand the revamping of the gravity irrigation systems to cater to the needs of non-paddy crops also.

Responses to the Failure

The above discussion reveals the growing mismatch between the canal irrigation systems and the demands of the changing agrarian scene of the State. It is reasonable to assume that this mismatch would have evoked responses from the farmers. They might have modified/improved existing rules or crafted new informal rules which defines their future interactions in realising irrigation. This could be viewed as a bottom-up evolution of irrigation institutions. Understanding these institutions could provide important insights into the constraints and factors that govern institutional choice. The analysis also helps in offering explanations for institutional inefficiencies and failure, and in providing remedial alternatives-based on the comparative study of institutions. In this backdrop, the next chapter discusses the theoretical foundations of institutions, existing empirical analysis and outlines an analytical framework for studying irrigation institutions.

NOTES

1. In this system water is delivered to farmers, below outlet through strict rotation schedules, in which the length of each turns in hours and minutes is proportional to the size of each farmer's holdings.
2. Under this system, every farmer is required to apply for irrigation every season indicating the crops to be irrigated and the water requirement. Once the irrigation authority approves the application, the agency is responsible to deliver water in amounts and on a schedule to bring the crops to maturity.
3. Irrigation water rights are assigned to land in this system, and lands are classified as entitled for specified number of crops. The irrigation authority is responsible to give water on an appropriate schedule to bring the crops to maturity.
4. Under *Satta* system farmers have to apply for water each season but there is no need to specify crops and rice is the assumed crop. The system was modified in certain regions to that of *assured irrigation* where every farmer is assumed to be a potential irrigator and are required to pay fee irrespective of the use.
5. *Economic Review*, State Planning Board, Government of Kerala (2001).
6. Net and Gross Irrigated Area in Kerala, 1960-61 to 1999-2000 (in ha.).

Year	*Net Irrigated Area*	*Gross Irrigated Area (GIA)*	*GIA as % of Gross Cropped Area*
1960-61	318711	456256	19.42
1970-71	431254	601385	20.50
1980-81	237970	381000	13.21
1990-91	317000	406000	13.69
1995-96	342193	435504	15.18
1999-2000	380043	470698	15.68

Source : Directorate of Economics and Statistics, Government of Kerala.

7. This order empowers the Government to direct every holder of land not to leave any land fallow, not to cultivate any other food crops than the one grown during the three years immediately before the commencement of the order or attempt to convert such land for any other purposes. Clause (3) of the order gives blanket powers to the Government which states that, 'notwithstanding any thing contained in any law, order, custom or practices for time being in force, if the State Government is satisfied that it is necessary or expedient to increase the production of food crops in any area,

they may, by order published in the Gazette, direct that every holder of land in that area shall grow, over such portion of his land and within such period as may be specified, in addition to any crop he may have grown over such land'. (*Kerala Land Utilisation Order*, Government of Kerala, 1967).

8. These factors are further elaborated in the subsequent discussions.
9. It is widely noted that the spread of education among labour and cultivating class created an aversion towards paddy cultivation and the physical labour involved in its cultivation.
10. The share of cultivators as main workers is only around 12 per cent as per the available estimates (*Economic Review*, Government of Kerala, 2001).
11. In the context of CRDS, since there had not been much time over-run, this issue do not invite a detail exploration.
12. District Collector chairs the advisory committee. The respective Executive Engineer, other relevant Government officials, local members of the legislative assembly, Members of Parliament and other non-official members are the other members, nominated by the Government.
13. This committee is to supplement the Irrigation Advisory Committee. The chair of the committee is the Junior Engineer responsible for the canal. Members include the President of the Panchayat , selected members of the Panchayat, and one representative farmer for every 400 hectares.
14. CADA provides for the constitution of Project Advisory Committee, Canal Committees and Water User Associations. The Project Advisory Committee include the following as members : one officer each from the Cooperative and Agricultural Departments, a representative from each Canal Committee in the scheme, relevant Members of Parliament, and Members of Legislative Assembly and five other nominated members. The Chairman is the District Collector and the Executive Engineer in-charge of the scheme is the convener. The primary function of the Committee is the seasonal planning, including making crop and water allocation plans. The Committee is supposed to meet at least one month before each season and monthly during the season. The Canal Committee constitute of the Presidents of Beneficiary Farmers' Association (BFA), and one officer each from the Cooperative and Agricultural Department, a representative of the Financing Bank and five nominated members. The Assistant Engineer in-charge of the Project is the convener of the Committee and the Chairman is selected from among the Presidents of BFAs. The Committee is supposed to meet at least one month before each season and once a month during the season. The function of the

Committee is to ensure equitable distribution of water, adoption of uniform agricultural practices in the command area, and coordination of the functions of the BFAs. Water Users Associations are the basic organisation under CADA. Each BFA has an elected Managing Committee consisting of seven members : President, Vice President, Secretary, Joint Secretary, Treasurer, and two Committee members. The term of the Committee is for one year from April 1—March 31. BFAs are responsible for water distribution and maintenance below the outlet. In addition, CADA channels subsidised inputs, including seeds and fertilisers for a variety of crops such as paddy, coconuts, pepper etc. through BFAs.

15. Water Rates in Kerala with effect from 1-7-74

Sl. No.	*Type of Land*	*Water Rates (Rs./ha)*
1.	Lands already registered as single crop wet lands and on which two paddy crops could be raised.	62
2.	Lands already registered as single crop wet lands and on which more than two paddy crops could be raised.	99
3.	Other lands already registered as wet lands on which two paddy crops could be raised	62
4.	Other lands already registered as wet lands on which more than two paddy crops could be raised	99
5.	Lands made fit for cultivation and on which only one paddy crop could be raised.	37
6.	Lands made fit for cultivation and on which two paddy crops could be raised.	62
7.	Lands made fit for cultivation and on which more than two paddy crops could be raised	99
8.	Other lands benefited	62

3

Institutional Choice

A Review of Theoretical Approaches and Empirical Evidence

The importance of institutions in influencing and determining economic outcomes has been a key concern of economic research in recent years. There has been an increased focus on the understanding of the micro-foundations of institutions and the process of institutional change. Application of institutional analysis has been significant not only within the study of firms or industrial organisations, but also in the analysis of agrarian systems and natural resource management. Irrigation institutions too attracted attention, in this context, during the past two decades.

Theoretical Considerations

Irrigation is widely considered as a subject of Common Pool Resource (CPR)[1] management in the theoretical discussions, due to its characteristics such as subtractability and costly excludability. In canal irrigation, the process of appropriation is by multiple users, simultaneously or sequentially, resulting in resource depletion (subtractability) and it is costly to exclude any appropriator from the system, whether or not they contribute to it or not. The theoretical discussions on common pool resource management largely confine to the possibilities/ difficulties in sustaining cooperation among self-interested agents in situations of strategic interdependence (Bardhan, 1993). The early characterisation of collective action is in the nature of a prisoner's dilemma game, in which defection or non-cooperation is the dominant strategy of each player, no matter what the other player does. Utility maximising individuals are assumed to choose to defect with the hope that if others co-operate, then he or she can free ride at others' expense[2].

Of late, considerable research has challenged the assumption of defection as a dominant strategy, even when the game is of a one-shot nature. Sen (1982) refers to the possibility of cooperation in cases of *prisoners' dilemma* situations, where either or both the players co-operate considering the loss in expected pay-offs when defection is the strategy. In a variety of situations individuals cooperate regardless of what other player does resulting in a *chicken game*[3]. Cooperation is also argued possible in other competitive situations as in the case of *assurance games*. Here, both the players prefer (cooperate, cooperate) to (defect, defect) as they are sure of an inferior equilibrium, in anticipation of similar counter strategy from the other player. Thus, even in cases of extreme *prisoners' dilemma* type situation, cooperative outcomes are possible under circumstances. Taylor (1987) argues that the constellation of costs and benefits of collective action on common pool resources is often of a kind, which is more favourable to the possibility of cooperation. To him, 'where individuals can choose from a continuous range of strategies their preferences are quite likely to be those of a *chicken game* or of a hybrid between a *chicken* and an *assurance game*'.

The possibilities of cooperation are also explained by using infinitely repeated games. This literature on repeated game theory shows that the long run interests of the foresighted self-interested individuals can sustain cooperative equilibrium. The proof of this has been provided by the *Folk Theorem*. The theory demonstrate that Pareto—inferior non-cooperative outcomes can be overcome, if repetition plays a disciplinary role for co-operation, as future behaviour is dependant on past actions. *Folk theorem* demonstrates that all outcomes including perfect cooperation and no co-operation may be sustained by repetition depending on the time horizon for which the game is played and discounting proposition. Thus, in an infinite horizon with undiscounted proposition, any pay-offs that is feasible and gives each player more than zero each can be sustained in an equilibrium. In a discounted formulation, *Folk Theorem*, for infinitely repeated game shows that any feasible discounted pay-offs that give each player, on a per period basis, more than the lowest pay-off that the player could guarantee in a single play of the simultaneous move component game can be sustained as the pay-offs of a sub-game Perfect Nash Equilibrium, if players discount the future to a sufficiently small degree[4].

For a finite but indefinite horizon, the theorem suggests that any feasible expected pay-offs can be sustained in equilibrium as long as each player has expected pay-offs at least as large as what that player can guarantee for himself/ herself even if all other players gang up against the player. That is, each player is told by others to stick to the agreement or everyone will gang up against that player. Then no single player acting alone has any incentive to deviate, which is the condition necessary for a Nash equilibrium. In the discounting formulation, pay-offs are discounted and seen. The possibility of co-operation in this case depends on the future discounted pay-offs, not being discounted too heavily, or the short-run rewards to defection (being not too large). If the benefits of defection equals or is somewhat equivalent to the discounted cost of defection in the form of credible punishment then co-operation is found to sustain.

Taylor (1987) argues that common pool resource interactions are mostly dynamic and involve interactions between individuals over time. In a dynamic cooperation thus, there is the possibility of conditional cooperation as there is interaction over time between different individuals choices. And the individuals' intertemporal preferences (how much he/she discounts future relative to present benefits) matter. Taylor suggests that if collective action problems are to be modelled as iterated games, then the appropriate model is a game iterated an indefinite number of times in which players discount future pay-offs. To Taylor, the individuals pay-offs (cost-benefit calculations), broadly depend on four components, such as : (a) the benefit from the increased amount of the good provided as a result of an additional contribution; (b) the cost of the contribution (both monetary and non-monetary); (c) the individuals' share in the cost of organisation; and (d) the economic and social benefits and/or costs which operates as selective incentives.

The proofs of the relevant theorems in the literature of discounted multi-person repeated games works on the basis of the possibility of administering sufficient punishments over time to outweigh the immediate benefits for the defector and of these punishments being credible. This, however, gives rise to a second order collective action problem, since punishment is costly to the punisher while benefits are distributed diffusely in the community. Strategies that punish players who fail to play their part in punishing the defector—meta punishment—is seen as the solution in this regard.

These theoretical models point to a potentially larger number of equilibrium outcomes, which will be chosen on the basis of observed past behaviour of others. Thus, even when costs and benefits are identical the degree of trust that the players have in others, play a crucial role. Seabright (1990) in his model of trust shows that many equilibrium points can exist depending on the players' belief on each other's trustworthiness. Successful experiences of co-operation in the past are identified to induce co-operation and its sustainability. Studies have shown that informal mechanisms may induce users of a common property resource to undertake collectively beneficial actions (Tang, 1993; Bardhan, 2000).

Apart from rationality behaviour based on pay-off calculations, moral and ethical considerations of the players are also argued to result in cooperative outcomes. Individual behaviour is ultimately a social matter and hence individual strategy would reflect a sense of identity involving recognition of other people's goals and the mutual interdependencies involved (Sen 1982, 1987).

The game theoretic models, thus, give insights into the possibility and sustainability of co-operation among self-interested individuals in a situation of strategic interdependence in the management of common pool resources such as irrigation. These models fail to explain the interactions among the agents involved in modifying the rules of the game. Thus, the approach is more or less static and can be used as a tool only to analyse strategic choices within an existing institutional set-up. However, different institutions can exist simultaneously and institutions can change over time. Thus, the crucial questions are how choices are made across institutions and what determines the change of institutions. Thus, it is important to understand the origin, the micro rationale behind the formation of economic institutions, the process of institutional change and factors, which help the formation of new institutions. The framework of new institutional economics provides the theoretical foundations for analysing institutional change over time.

The Framework of New Institutional Economics

Of late, the institutional approach based on transaction costs, has become an important framework in the understanding of institutions. The framework, originally developed to explain the nature of the firm or organisation, has been extended in analysing institutions in agriculture and common pool resource management. Institutions are defined as a set of formal or informal rules/norms, which govern the behaviour of individual and/or groups. It is thus, the rules of the games in a society, or more formally, the humanly devised constraints that shape human interaction (North, 1990). The central concept of this framework is the existence of positive transaction cost as against the conventional neo-classical assumption of costless transaction. Transaction costs are the costs of measuring the valuable attributes of what is being exchanged and the costs of protecting rights and policing and enforcing agreements (North, 1990 : 27). These include *ex ante* costs of negotiating and forming a contract and the *ex post* cost of monitoring and enforcing it. When it is costly to transact, institutions matter (Coase 1960).

A major stream of institutional analysis is the property rights school, which is mainly associated with the works of Coase (1960), Demsetz (1967), Alchian & Demsetz (1972, 1973) and Posner (1977). The conceptual foundations of the school were laid down by Demsetz (1967) in the article, *'Toward a theory of property rights'* as :

> "Changes in knowledge result in changes in production functions, market values and aspirations ... the emergence of new property rights takes place in response to the desires of the interacting persons for adjustment to new benefit-cost possibilities ... property rights develop to internalize externalities when the gains of internalisation becomes larger than the cost of internalization".
>
> (*Demsetz*, 1967 : 350)

This literature emphasises that property right structure has a direct bearing on transaction costs. Property right involves three aspects : right to use the asset, the right to appropriate returns from the asset, and the right to change the form and/or substance of an asset. Depending upon the definition of these three aspects, property right structure changes and has direct bearing on transaction costs.

North (1990) uses a variant of the property rights model to explain institutional change where also transaction cost is taken as the key factor. Transaction costs arise because of the costs of measuring multiple valuable dimensions involved in exchange and because of the costs of enforcing agreements. As per this model, institutions evolve in response to transaction costs as individuals or groups innovate institutions to reduce transaction costs.

> "Economic institutions are innovated because it appears profitable for individuals or groups in society to undertake the costs of bringing about such changes. The

purpose is to capture for the innovator some profit unattainable under the old arrangement. The essential requirement for initiating an institution is that the discounted expected gains exceed the expected costs of the undertaking".

(*North and Thomas,* 1970 : 3)

Efficient institutions replace old and inefficient ones if the net gains of institutional changes are positive. Changes in relative prices, which affect transaction costs, are the main force behind institutional change in this school of thought.

In the transaction cost framework, the role of information costs is important both in direct cost terms as well in their impact on the economic behaviour of actors. The costliness of information is the key to costs of transacting, which consists of the costs of measuring the valuable attributes of what is being exchanged and the costs of protecting rights and policing and enforcing agreements (North, 1990). Further, information is not only costly but incomplete, and enforcement is not only costly but imperfect (North, 1993). This could result in the existence of inefficient institutions[5]. Further, even if the institution was efficient at one point of time, it need not be the optimal one at a different point of time. Inefficient institutions are presumed to persist and could continue to influence the choices made in future, which is attributed mainly to the self-reinforcing feed-back provided by the organisations developed within the institutional framework (North, 1990). This inertia towards institutional change is referred to as, *institutional lock-in or path dependence* (Arthur, 1989; David 1985; North, 1990).

Institutional approach helps in understanding the evolution of institutions and institutional change through the analysis of transaction costs. Though the framework of new institutional economics has been widely accepted as an important tool in understanding and analyzing common pool resource management not much empirical works are available within this framework.

Empirical Studies on Irrigation Institutions

An earlier set of empirical studies on irrigation focussed on the performance of irrigation management by the State. The relationship between irrigation and general political authority have attracted much attention from historians and social scientists since long. Most of the studies on the efficiency aspects of the State controlled canal irrigation system highlight the flaws of bureaucratic, authoritarian governments. The influential role of politicians and the scope for rent seeking by agency staff responsible for allocating and delivering water under public provision has been widely discussed (Wade, 1982; Repetto, 1986). On the basis of a field study of State-constructed, operated and maintained irrigation system in a South Indian State, Wade (1975) views administrative and political corruption as the important reason for the poor performance of canal irrigation. Pant (1981) discussing the Kosi CADA Project in Bihar observes the neglect of institutional considerations such as operational procedures of irrigation bureaucracy and the over-emphasis on technical diagnosis as the basic problems with government initiated collective action. The lack of co-ordination between administrative decisions of the State and actual requirements of village communities has found to result in under-utilisation of resources and socially non-viable outcomes. While explaining the case of Tanrwan village served by Sone canal system in Bihar, Sengupta (1991) cites the case of imposing a pre-established organisational form on an already functioning informal farmers' organisation leading to a mismatch between the formal design and informal rules.

Using data from Gujarat, Jayaraman (1981) points out another aspect of improper canal administration, with excessive spending on salary component and inadequate allocation for physical repairs (operation and maintenance). Brewer *et al.* (1999b) analyses the mismatch between the rules defining delivery pattern and the actual performance. He has also noted that the actual delivery do not meet the requirements of water users. Using data from the Tambraparni irrigation system in Tamil Nadu, it is argued that inconsistency in the water distribution rules creates difficulties in system operations that are likely to lead to inefficient and inequitable water distribution.

Bandaragoda (1998) focuses on the existing gap between the original design of *Warbandi*[6] as a water allocation method and the way it is practised among irrigators. A study of 22 tertiary level irrigation sub-systems (water courses) located in 6 secondary canals in 3 different major canal systems of Pakistan's Punjab province showed that, the actual practise of *Warabandi* water allocation schedule differed substantially from the design to the extent that 'not a single water course in the study region adhered to in daily water distribution operations'.

The lack of an appropriate policy of pricing and its recovery has also been a major area of discussion. In a survey of 17 irrigation projects (Duane, 1975), it was found that only 30 per cent of the total project costs were recovered. In many cases, this amount recovered do not even meet the O&M charges leading to the deteriorated performance of the system. The pricing of irrigation as an economic good is seen as an important step towards increasing the efficiency of irrigation water use (Sampath, 1992; Perry, 2001). The studies on the whole reveal that policy issues and the associated administrative system are central in the success or failure of irrigation institutions, especially in centralised public provision, where state assumes the prime role.

Empirical Studies using New Institutional Approach

There have been a number of studies during the last decade, which have attempted to use the insights of the new institutional approach to analyse irrigation institutions. These studies have identified and delineated a number of factors determining the existence of particular institutions; conditions under which co-operation fail or succeed; and the equity and efficiency aspects of various institutions. In the study of irrigation institutions, a major stream of research is centered around the issue of co-operation building among irrigators (Wade, 1975, 1982, 1988a, 1988b; Jayaraman, 1981, Oakerson, 1986; Ostrom and Gardner, 1993; Ostrom, 1995).

The major empirical model towards the study of common pool resource management within the institutional approach has been the framework, developed as part of the Workshop in Political Theory and Policy Analysis, and associated with Kiser and Ostrom (1982), Oakerson (1986) and Ostrom (1986). In this framework, four different factors are basic in analysing the working of various institutions of natural resource management—physical and technical attributes, decision-making arrangements, patterns of interaction and outcomes

In the framework, the physical characteristics of the resource identify features of jointness, exclusion and indivisibility. Decision-making arrangements reflect the operational rules, which specifies the user behaviour as to who makes decision and what decisions are made. The definition of property right is the crucial variable affecting the attribute. Together the physical and technical attributes as well as decision-making rules determine the patterns of interaction with regard to the resource and hence the outcome. In the short run the physical, technical and decision-making attributes are taken as fixed or unchanging, while in the long run they change leading to institutional change. Individual's responses to outcomes (efforts to improve outcomes) are also taken as an important variable in institutional change.

The framework has also been applied in the understanding of irrigation institutions by Ostrom (1990), as *Institutional Analysis and Development (IAD) Framework*. The IAD framework, based on the transaction cost analysis, is concerned with identifying appropriate institutional arrangements that can counter perverse incentives in various transaction situations (Tang, 1992). The focal point of IAD framework is the action situation in which individuals adopt actions or strategies (the presence or absence of cooperative behaviour), structured by three sets of contextual attributes in an irrigation system namely; the physical attributes of the irrigation system, the attributes of the community of participants and the set of institutional arrangements in use by participants. The central question addressed under this framework is what are the incentives for individuals in a setting that lead to act in ways that generate patterns of outcomes (Lam *et al.*, 1997). Incentives are central in the model and accordingly, different operational and collective action rules, in combination with the physical and community attributes of an irrigation system, can create different incentive structures that induce co-operation or conflict among individuals. This model, however, does not provide any fixed theory in explaining irrigation institutions and its change rather gives only a set of tools that explains different situations in the management of irrigation. Thus, it is useful only in situational analysis and in the understanding of sustainability of irrigation institutions.

Kolavalli examines the conditions of collective action and user incentives and provides useful empirical insights into the importance of transaction costs in shaping farmer decisions on participation in local institutions (Kolavalli, 1995, 1996). The studies highlight that, apart from the fixed organisational costs, collective action involves costs that are incurred to obtain agreements on sharing of costs and benefits and to coordinate and organise the collective effort. Based on the empirical evidence it is shown that the individuals selects a particular set of decision-making rules among the alternatives, which minimises the expected interdependence cost of organising. Sengupta (2001) attempts a detailed enquiry of irrigation institutions in Magadh (Bihar) within the theoretical approach of new institutional economics, where individual actions are captured within the bounds of institutional settings. Based on historical data, the study analyses the dynamics of uncertainties and self-organisation process in the three different phases of evolution, stability and transition of irrigation institutions. Failures of local institutions are conceived in terms of conflicts with the imposed legal system, which neglects traditional structures. As a result, it is held that, the state of affairs is one of institutional disorder, resulting in poor performance.

A major chunk of empirical studies on irrigation institutions focuses on the conditions or factors that facilitate or determine cooperation among farmers. The findings of these studies could provide valuable insights into the functioning of collective action institutions and in understanding the shaping and crafting of institutions, which in turn help in delineating the determinants of institutional choice in irrigation. The conditions or factors highlighted in the empirical studies (field as well as experimental) that are important in the functioning of institutions can be broadly divided into three categories : (1) Locational Factors; (2) Land and Crop Characteristics; and (3) Group and Farmer Characteristics.

Locational Factors

Locational factors are identified as an important in influencing the irrigation institution prevailing. Farmers' vulnerability to scarcity and uncertainty in water supply due to locational disadvantages and its effects on their incentive to co-operate have drawn special attention. Wickham and Valera (1979) in a study of irrigation projects in Philippines, observe that farmers have less incentive to organise if they do not have predictable or sufficient flow of water into their watercourses. The asymmetries that exist between headenders and tailenders on irrigation systems are extreme and found to create conflict and lack of co-operation among farmers in general (Lam *et al.*, 1997). Nevertheless, scarcity and uncertainty situations is also argued to induce disadvantageously placed farmers to struggle for obtaining more water leading to increased co-operation among irrigators in a particular locality. Wade (1988a), drawing upon the experiences in South India, argues that greater the scarcity and uncertainty of water supply, the greater the likelihood that a community of cultivators will develop collective action arrangements to govern their watercourse. However, Bardhan (2000), in a field study of 48 villages cautions of the possibility of failure of co-operation among tailenders where extreme scarcity exists. Wade (1987) provides empirical evidence to show that the distribution of control over land in relation to proximity to the outlets strongly influence the nature of collective response to a given water scarcity situation. This is also confirmed by Boyce (1987) while explaining the case of tubewell distribution in Pakistan and by Bandopadhyay and Von Eschen (1988) in the context of irrigation in Bengal. These contradictory evidences have been explained in a general context where irrigator's vulnerability to scarcity and uncertainty in water supply is related in a curvilinear manner to their incentives to cooperate. This is indicative of the condition that farmers have to be sure of at least some minimal availability of water before they are willing to invest in collective action.

Crop Characteristics

Functioning of institution is also found to be influenced by the land use and crop pattern, as farmers' participation in water transactions directly depend on the needs to control water. These needs are directly related to the cropping pattern and crop water requirements (Rinaudo *et al.*, 1997). The prevalence of co-operation is reported higher among paddy and wheat farmers, which require large quantities of water during specific seasons. Shah (1993) shows that farmers depending on water market arrangements mostly cultivate high profit crops requiring protective irrigation. Increased cost of getting water leads to increased production of crops requiring less water and reduced acreage of crops requiring high volume of water. The establishment of tradable property rights through assigning high value to water also would lead to the use of scarce water in the irrigation of high valued crops (Rosegrant and Binswanger, 1994).

Individual and Group Characteristics of Farmers

Group characteristics such as number of users, homogeneity in terms of ethnic, social, cultural and economic status, presence of leadership qualities, personal characteristics such as occupation, age and experience are found important in empirical studies in influencing the existence and sustainability of collective action. It has been found that other things being equal, it will be easier to organise collective action in irrigation system of smaller sizes with fewer users[7]. This is related to the cost of information gathering, communication, decision-making and monitoring costs which are found to be high when the size of the group is large. Bardhan (1993) observes that co-operation was better in small groups with similarity of hydrology, water requirements and clearly defined boundaries. In his two empirical studies on Gujarat, Jeyaraman (1981) also endorses the viability of small group in bringing co-operation. He finds that a core committee of small groups could also ensure co-operation even when the group is large.

Coming to the intra-group characteristics, studies have established the importance of homogeneity of irrigators in terms of socio-economic conditions, payoffs and distribution of plots, crop and farm characteristics. Irrigators differ from one another in their cultural and social characteristics such as ethnicity, caste and religion. These differences are important contextual attributes that affect the institutional choice of irrigators (Bardhan, 1993). Bardhan (2000), in his study of irrigation communities in South India, establishes that social homogeneity is conducive to co-operation. The Gini co-efficient for the community with 75 per cent or more of the farmers belonging to the same caste groups was found positive, clearly confirming the role of group homogeneity as a condition for collective action and cooperation.

Homogeneity in income characteristics also assumes importance in cooperation building. From a study of 23 community irrigation systems, Tang (1993) finds that lower the variance of average family income among irrigators, higher the degree of rule conformance and good maintenance. Jeyaraman (1981) also notes the importance of relatively egalitarian structure of the community denoted by size of land holding and 'equal' interests (for instance all the group members are full time cultivators), in farmers co-operation in surface irrigation projects in Gujarat. Based on an action research programme conducted at 4 pilot sites in Pakistan, Bandaragoda (1999) attributes the success of farmer participation in management to the homogeneity of the group, with a majority of medium and small landowners.

Presence of experienced hands (in terms of age and training) in leadership and importance of leader being within the community—who is also an irrigator and share the same egalitarian structure and interests with other irrigators—is also found crucial in determining co-operation. (Jeyaraman, 1981). Bandaragoda (1999) confirms this finding while highlighting the role of 'fairly established' rural leadership as the driving force for concerted action. Meinzen-Dick *et al.* (2002) also highlights the importance of group characteristics such as leadership and social capital (indicated by influential persons, college graduates, etc.) in the successful functioning of user associations in the context of major canal irrigation systems in Karnataka and Rajasthan.

Rules, regulations and penalties followed in the groups are also widely captured in the empirical studies. Citing the example of farmers from Shivalik range of Himalayas, Chopra, *et al.*, (1989, 1990) stress that equal distribution of pay-offs and definite and unambiguous norms and rules facilitate co-operation. This would facilitate the observance of rules of water distribution in terms of water rights, number of discharges and water turns. Jeyaraman (1981), based on the empirical studies in Gujarat shows how rules of certainty in irrigation supplies from gated turnouts facilitated group action among the farmers. The knowledge that pursuit of short-term interests may harm long-term aims by affecting the reactions of other farmers in future interactions is cited as a powerful inducement to co-operative behaviour by many studies. (Runge, 1986; Sugden, 1986; Wade, 1982, 1988b). Also, the introduction of penalties on violation of agreements in co-operative action was found important for sustaining co-operative behaviour. Involvement of the users in the design and enforcement of distribution rules are also cited as crucial in the success of co-operation (Brewer *et al.*, 1999b).

On the whole, as evident from the foregoing discussion the empirical studies on irrigation institutions focus largely on explaining the incentives and constraints that induce individuals to seek cooperative outcomes. The empirical verifications are, thus, mostly confined to the explanations of existing institutions, their failure and success, and the conditions under which various institutions develop and sustain.

An Analytical Framework for the Study

Irrigation institution can be perceived as a set of working rules in the provision and use of irrigation water in a particular location to reduce uncertainty and conflict among irrigators. These institutions reflect the choice domain of independent economic actors in the *status quo* and define the interactions among irrigators. Multiple institutions exist due to varied set of incentives and constraints, faced by individuals. The coexistence of multiple institutions raises the question as to how choices are made across institutions and what determines institutional change. The framework of New Institutional Economics provides the theoretical foundations to address this. Under this approach, the choice of institutions and institutional changes are analysed through the economic manifestations of changed circumstances in terms of costs (including transaction costs), prices and profits.

Changes in the *status quo,* in terms of constraints and possibilities induce individuals to strive for improved institutions. In response to the new conditions, individuals undertake efforts to modify the existing institution so as to bring them in line with the new scarcities, technological opportunities, new distributions of income or wealth, or the new tastes and preferences. Thus, crafting of new institutions is the manifestation of rational choice of individuals. Emergence of new irrigation institution reveals the evolutionary process by which the continued interaction of individuals results in changed rules and norms for appropriating water.

Farmer's choice of institutions is also influenced by the policy environment. This includes the state policy governing irrigation management that determines the priorities of projects, boundary between market and non-market processes and allocation of net benefits, etc. The macro decisions as to whether provision of irrigation water is a state responsibility or whether it has to be subsidised and so on are all decided at the policy level.

The repetitive action and interaction of individuals lead to institutional change under two circumstances. Firstly, it may result in collective response leading to changes in the institutions at the policy level. Secondly, there can be institutional change at the local level in response to the changes/*status quo* at the policy level. This effort to change institutions can also include attempts to subvert existing institutions. For example, when, the defined rules of a pattern of water delivery (irrigation institution) do not match the goals of water users, the users will subvert the rules individually. These sub-versions in practice after repeated trails assume the shape of new institutions, taking the individual farmer closer to equilibrium—at least until the next exogenous change stimulate further economic response.

Analysis of Farmer Responses

If water is abundant, additional cost of appropriation of irrigation will be zero to all farmers. In situations of scarcity, farmers do incur costs to ensure reliable supplies. However, each farmer tries to minimise the costs involved for a particular level of water availability and control. This in turn is an act of maximising the net pay-offs. This optimisation behaviour of farmers is reflected in the institutional choice. One way of cost minimisation is the alteration or modification of the existing institution. Farmers effect changes within the structure of the institution, by redefining the norms, rules and rights (entitlements). The other way is to devise alternative institutions, with visible structural changes in terms of rules, claims and entitlements. In either way, the optimising or rational behaviour of the farmers is influenced by a host of constraints. In the case of irrigation, it is reasonable to categorise these factors into three sets such as locational, land and crop specific, and individual or household characteristics. The relative influence of these factors in the institutional choice is, however, expected to vary widely across plots and individuals.

The co-existence of multiple institutions in a command area, with different cost implications can thus be explained as a constrained optimisation exercise, based on rational behaviour of farmers. The existence of distinct sets of constraints (locational, land and household characteristics) results in varied pay-offs from the same institution for different farmers. Thus, an institution that is efficient (in terms of higher pay-off) to a farmer need not be so to another, due to the difference in the set of constraints. This also explains the situation of same farmer resorting to alternative institutions in different plots.

Each farmer chooses between different institutions for irrigating the plot with different cost and benefit implications, within the constraints. A rational farmer selects a particular institution that minimises the additional cost, thereby maximising the net pay-offs, given the benefits. As benefits from farming largely depends on the profitability of crop under cultivation, given the crop mix, the rational behaviour of farmers could be deduced to that of cost minimisation problem. In this case, farmers choose a particular institution, which ensures the desired level of water at the lowest cost. The costs associated with various institutions can be broadly divided into two : production and transaction costs. With positive transaction costs associated with various institutional alternatives of acquiring irrigation water, economising takes place not with reference to production costs but with total costs, which is the sum of production and transaction costs. A farmer selects a particular institution that minimises the transaction cost (given production costs) thereby maximising the net pay-offs.

The choice of a particular institution at a point of time does not mean that the farmer will always stick on to that institution. So long as there exists positive transaction costs in appropriating water, the farmers do have incentive to strive for alternatives that assure reduction in cost. This logic could explain the varied behaviours of farmers in institutional choice such as alteration of existing institution, devising of new institution, reverting to an earlier institution, or giving up the irrigation intensive agricultural activity partially or fully[8].

Institutional change, thus, is driven by economic calculations of individuals or group perceptions about the *status quo, vis-a-vis* some alternative state in which things are presumably better. This perception of change is induced by individual expectations of pay-offs, and not by aggregate gains to the society. Thus, the choices that reveal economic preferences of individuals may not be socially efficient.

Institutional efficiency can be analysed in terms of, (a) changes in productive efficiency, and (b) redistribution of income (Bromley, 1989)[9]. Changes in *productive efficiency* refer to institutional changes that lead to an indisputable increase in the net social dividend macro-policy. Here social efficiency is not deteriorated if, the gainers could compensate the costs of the worse offs. In this case, institutional change brings in benefits and leads to an upward shift in social welfare function. This change thus entails increase in both individual and social dividend and thus efficiency. Certain institutional changes do not entail increased social dividend, but only leads to *redistribution of income*. Redistribution of income would mean an improvement in the economic position of some individuals, without any change in the aggregate production in the society. Here, the institutional choice results in skewed distribution of benefits (in terms of income) in favour of certain individuals.

The foregoing discussion highlights the possibility of examining the co-existence of multiple institutions in irrigation within the framework of new institutional economics. The study uses these insights to analyse the institutional choice of farmers by a case study in the command area of Chalakuddy River Diversion Scheme in Kerala.

NOTES

1. Common pool resources are natural or man-made resources that are large that it is costly (but not impossible) to exclude potential beneficiaries from obtaining benefits from their uses (Ostrom, 1990).
2. Two seminal writings on Common Pool Resources in 1960s, *'The logic of collective action'* (Olson, 1965) and *'Tragedy of Commons'* (Hardin, 1968) provided the basis of the discussion.
3. In a *Chicken game* at least one of the players would cooperate as the consequence of defection is so bad leading to losses in net-pay-offs of both the players.
4. For details of Folk's theorem, see Mas-Collel, Whinston and Green, 1995.
5. For a discussion of institutional efficiency *see* North (1990) and Rutherford (1994).
6. *Warabandi* is a rotational method for equitable allocation of available water in an irrigation system, by turns fixed according to a time roster, specifying the day, time and duration of supply to each irrigator.
7. The early work of Olson also suggested that collective action work better in small groups (Olson, 1965). Models of repeated games also suggest that cooperative strategies are more likely to emerge and be sustained in small groups.
8. The shift towards less irrigation intensive crops, keeping fallow lands, conversion of agricultural lands to non-agricultural purposes, disposing of farming fields etc. come under this category.
9. For detailed discussion of these various concepts *see* Bromely (1989).

4

Irrigation Institutions in the Canal Command of CRDS

As discussed in the previous chapter, irrigation institutions could be defined as a set of working rules in the provision and use of irrigation water. These institutions reflect the choice domain of independent economic actors at a given point of time. An understanding of the co-existence of various irrigation institutions warrants a detailed examination of the characteristics and functioning of each institution. Accordingly, the subsequent discussion in this chapter focuses on different institutions that prevail in the distribution of irrigation in the command area of Chalakkudy River Diversion Scheme (CRDS).

Profile of Chalakkudy River Diversion Scheme (CRDS)

Chalakkudy River Diversion Scheme (CRDS) was launched under the First Five Year Plan with an objective of increasing paddy production in the State. Partially commissioned in 1952 and completed in 1957, the scheme consists of a diversion weir across Chalakkudy river, at Thumburmuzhi in Thrissur district (Figure 4.1), having a length of 185 metres and height of 4.11 metres. There is no independent storage for CRDS and the source of water is the tail race and surplus water from the Peringalkuthu Hydro-Electric Scheme and the drainage water from the catchment of Chalakkudy river below Peringalkuthu.

The Scheme has two major canals—the Left Bank Canal (LBC) and Right Bank Canal (RBC). The total length of the main canal is 100.258 km, of which RBMC has a length of 48.058 km and LBMC of 52.200 km, with designed capacity of 12.74 m^3/sec. The length of the branches under RBMC and LBMC are 111.356 km and 108.726 km, with distributaries of 22.684 km and 3.776 km respectively. The total length of the canals is 345.32 km and bed width off-take is 6.10 m^1. The canal is unlined in most of the reaches. Spouts are provided in the branch canal/distributaries and the farmers make their own arrangements to carry water from the spout to the destination at their cost and convenience. Being linked to a hydel project, water is let down in the canals throughout the year except in April and May, which are left for maintenance work.

The canals have been designed to cater to an ayacut (command area) of 19,690 hectares of paddy land in Mukundapuram, Kunnthunadu and Parur Taluks of Thrissur and Ernakulam districts (Figure 4.1).

However, according to a verification conducted by PWD and Revenue Department in 1984, the area actually irrigated was 13,530 hectares. Water is supplied to the field for the first crop (*Viruppu*) during June to September, for second crop (*Mundakan)* during September to December and for third crop (*Puncha*) during January to April. The *Puncha* ayacut as per the project planning is less than the other two. At present, the

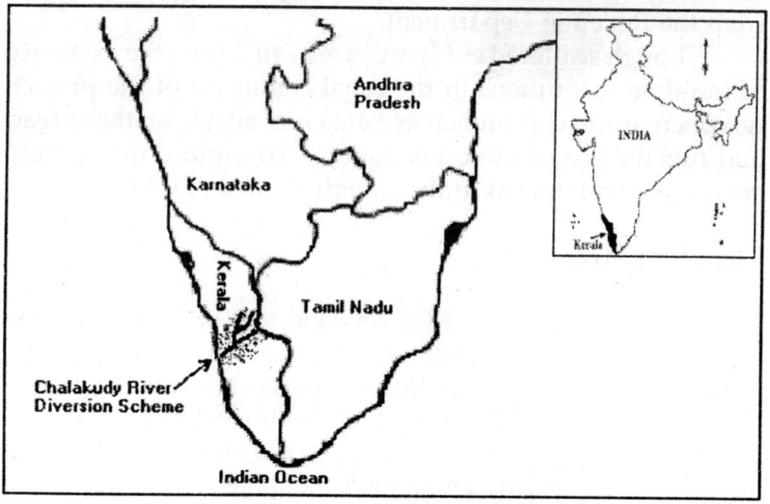

Figure 4.1 : Location of Chalakkudy River Diversion Scheme

entire area of 13,530 hectares is provided with irrigation during the first and second crop seasons. During the third crop season, irrigation is provided to only 6,925 hectares, which vary from year to year.

The CRDS is under the administrative control of Executive Engineer, Irrigation Division, Thrissur who is assisted by two Assistant Executive Engineers and nine Assistant Engineers in running the distribution system as well as accounting and administration matters. As per the initial design, the entire ayacut is irrigated simultaneously. However, the bad condition of the canal necessitated the turn system. Unlike in most of the other irrigation projects, in Chalakudy, there are no field *boothies*. Spouts are provided in the branch canals/distributaries and the farmers make their own arrangement to carry water from the spout to the field at their cost and convenience and no records are maintained regarding the supply of water to the farmers. There are no permanent water-measuring devices to assess the quantity of water. The Revenue Department is in-charge of collection of water cess, which is fixed at Rs. 40 for one crop and Rs. 60 for two crops annually per acre. The cess is not collected regularly; and often it along with arrears are charged at times when farmers require any other service from the Revenue Department.

Though scattered evidences are available on the existence of multiple institutions in the canal command of the project, no concrete and comprehensive data is available on the spread and functioning of these institutions. To capture this, a field survey was conducted in the command area of CRDS.

The Study Area

The Left Bank Canal of CRDS was selected for conducting the study. This selection was done on the basis of the findings of a pilot survey carried out in different parts of the ayacut, which revealed that the characteristics of both the bank canals are more or less similar. The particulars of the Left Bank Main Canal are given in Appendix 4.1 and 4.2.

The left bank canal system consists of 31 branches and 2 distributories. According to the latest available estimate of ayacut, the Left Bank Canal has a total ayacut of 7338.14 hectares, of which 2521.63 hectares is under *Virippu*, 1743.74 hectares under *Mundakan*, and 3072.77 hectares under *Puncha*. The canal irrigates 5 villages in Mukundapuram and 6 villages in Aluva Taluks.

Selection of Sample Plots

The unit of analysis and sample plots were selected keeping in view the technical specifics of the canal system. A canal system is fundamentally a mechanism for conveying water from its source to the fields through a network of channels, wherein the flow of water is based on the gravitational pull. The location of the plot is an important variable in the canal network, which influences the availability of water, as is evident from the Figure below (Figure 4.2)[2].

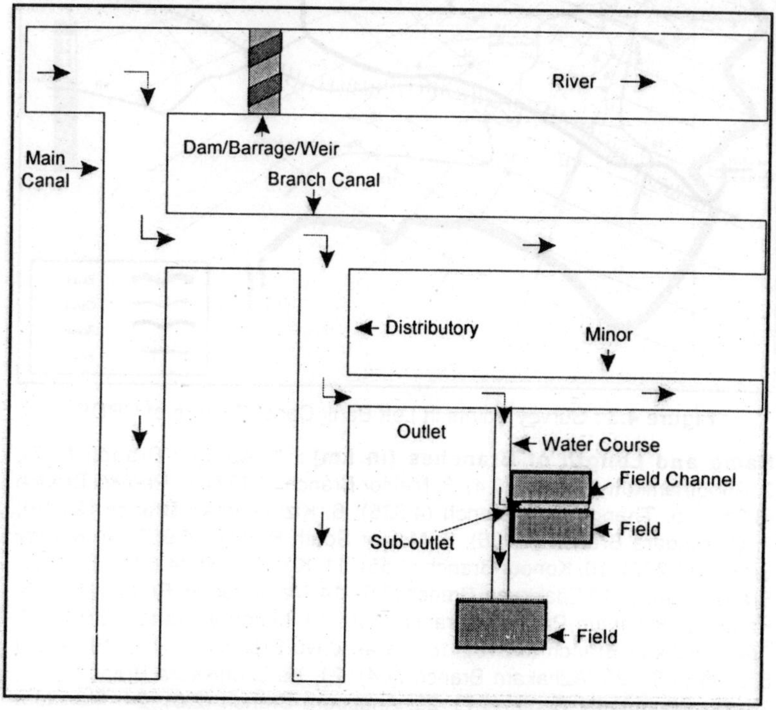

Figure 4.2 : Layout of a Typical Canal Network

To adequately represent the various heads a multi-stage stratified sampling was resorted to. The entire command area is divided into different geographical trajectories. The parameters for stratification and division are the distance from main canal, branch canal, and outlet, elevation of the plot. To give representation to different locations; head middle and tail end, trajectories were drawn both vertically and horizontally on the map of the project. A diagrammatic representation of the same is given in Figure 4.3.

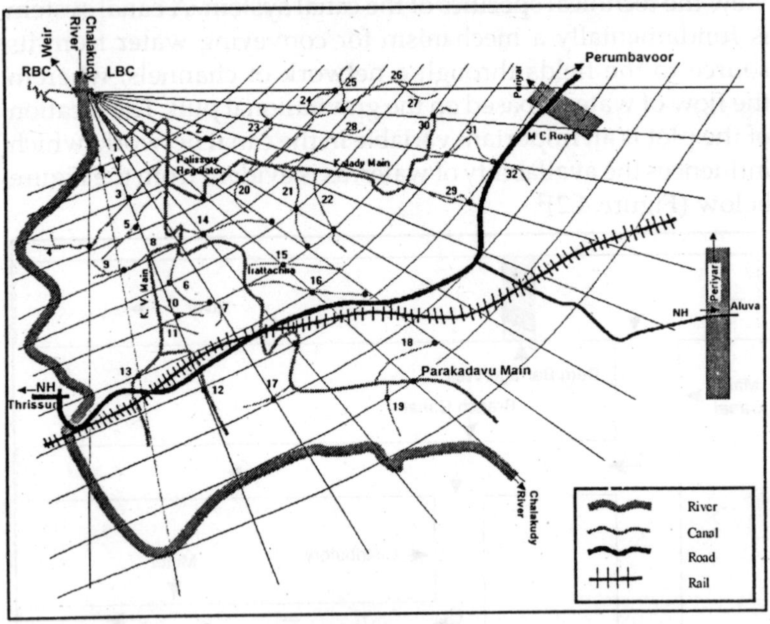

Figure 4.3 : Survey Points in Left Bank Canal System of CRDS

Name and Length of Branches (in km) : 1. Adichili Branch (7.7); 2. Bhoothamkutty Branch.(6.4); 3. Meloor Branch (6.437); 4. Poolani Branch (3.218); 5. Thanguchira Branch (4.525); 6. Kizhakumari Branch (2.816); 7. Chirangara Branch (1.006); 8. Meloor South Branch (4.968); 9. Koratty Branch (1.207); 10. Konoor Branch (1.55); 11. K.V. Branch (4.627); 12. M. V. Branch (4.425); 13 Edakkunni Branch (3.6); 14. Marangadam Branch (1.4); 15. Karukutty Karayam Parambu Branch (6.2); 16. Mambra Branch (2.97); 17. Peechanikkad Branch (3.219); 18. Parakadavu Branch (3.319); 19. Attara Branch (1.8); 20. Azhakam Branch (1.4); 21. Karukuttikkara Branch (2.6); 22. Bhoothamkutty South (1.7); 23. Anappara Branch (2.6); 24. Manjapra Branch (8.6); 25. Naduvattom Branch (2.339); 26. Komarapadom Branch (4.00); 27. Vengoor Branch (6.0); 28.Thottakam Branch (2.6); 29. Kalady Branch (5.2); 30. Chengamanad Branch (1.6).

The points of intersections in the figure were selected as sample points. Accordingly, 35 intersection points were selected on 29 branches and the main canal. From each point, 6 households were selected randomly. Thus the total number of sample households were limited to 210. The unit of analysis is individual plots rather than individual farmer, considering the possibility that a farmer can have multiple plots with different institutions. Thus, if a particular farmer possesses more than one plot, each plot was treated as a separate sample. For the 210 households surveyed the number of sample plots were 474. Out of these, 77 were unirrigated plots. The inclusion of these plots was due to the fact that they are in the ayacut though not irrigating. With the exclusion of these, the study focuses on the 397 plots, which are irrigated.

Information was collected through a structured questionnaire through which, data regarding area owned, area irrigated, water adequacy in the canal, distance to main canal, branch canal and outlet, elevation of the plot, land and crop details, sources of irrigation, number of intervening farmers, time taken to irrigate the crops, stress days, distribution of wells, yield, credit and input availability, input-output prices, personal information of the households, water charges, and additional cost spend on alternative systems of water distribution both in terms of labour and money cost were collected from the farmers. The results of the primary survey were supplemented by discussions with farmers, farmer groups, local body members, irrigation department officials and other key persons. Detailed case studies of various institutions were also carried out.

Failure of Canal System in CRDS

The emergence of alternative irrigation institutions within the canal command can be seen as an outcome of the inadequacy of the major irrigation project. To understand the evolution and functioning of alternative institutions, it is imperative to gain insights into the reasons for the failure of the CRDS canal system. With this objective, the field survey tried to capture some of the critical aspects of performance of the CRDS.

During the last two decades, it is widely admitted that the institution of public provision has failed in providing adequate and timely water according to the requirements of the farmers in the ayacut of CRDS. The reasons for the failure of the canal system can be broadly discussed under two heads : (1) Inadequate and uncertain irrigation; and (2) Changed irrigation requirement.

Inadequate and Uncertain Irrigation

In CRDS, the farmers' access to irrigation is in terms of allocation of water per unit of land. Supplies to water-courses are proportional to their command area. Within the water-course, canal water is distributed to farmers according to a rotational schedule. However, it was reported that these turns are often not followed. From the survey it was evident that during periods of peak irrigation requirement, the supply becomes potentially insecure and inadequate in the command area due to a multitude of factors.

One of the major reasons for the water scarcity in CRDS canals is their inadequate operation and maintenance. As a major portion of the canal network is earthen and unlined, the lack of proper maintenance has resulted in silting and damages in the canal network, leading to irregular and deficient availability of water. According to an official report of the Irrigation Department, out of the revised total ayacut of 13,530 hectares, only 7,417 hectares (3,150 hectares in RBC and 4,267 hectares in LBC) could be fed on account of poor O&M (Government of Kerala, 1994). Thus, non-timely and insufficient provision of maintenance grant and lack of interest on the part of contractors to take up small works are the reasons cited by the irrigation authorities for the reduced availability of irrigation water. Apart from this, farmers believe that, political interference and corruption in granting contracts also adversely affected the operation and maintenance of the canal network.

Further, it was observed that, despite water scarcity, the command area is reported to have got extended due to political considerations and nepotism. Extension of canal is often used as a strategy by political leaders in the area to appease the voters. Some reported cases of political interventions are : (a) Extension of irrigation to areas, where it is required to be released only for drinking and other domestic purposes; (2) Provision of water for *puncha* (third crop) even in some branches where farmers are expected to get water only for irrigating second crop, as per the original plan; (3) Extension of irrigation to areas which do not come under the *ayacut* as per the design through the construction of unauthorised distributaries or outlets[3]. There are some instances of local leaders seeking to fulfil promises made by them in their election manifesto. Significantly, farmers in the location recollect that most of the major extensions in the canal network coincided with the elections to the Parliament or Legislative Assembly.

Administrative mismanagement has also contributed to the uncertainty in water availability. This invariably leads to failure in adhering to the time-table for regulating water to different branches. This is partly due to the reduced availability of water on account of the poor O&M. Added to this, the lack of commitment, corruption and other administrative and political mismanagement, which have further worsened the situation. The respondents recall bribing of canal lascars and other irrigation officials in earlier times to bring water to their branches and outlets. However, of late, there is a reported decline cases of bribing and corruption[4]. Political intervention in diverting water to specific branch canals was also noted[5].

The reduced availability of water for diversion is also to some extent due to the reduction in the release of water from the Peringalkuthu Dam of the Kerala State Electricity Board (KSEB). The storage capacity of the dam is reported to have reduced because of sedimentation as a result of deforestation. Also, the objectives of Irrigation Department and the Kerala State Electricity Board are found to be conflicting. KSEB is reluctant to release more water during summer when irrigation demand is high since its priority is to preserve water for power[6].

The other important factor that has reduced water availability in the tail ends is the reported overuse of water by head end irrigators. During the survey, it was observed that flooding of plots was a common practice followed by head reach farmers. Unscientific methods of irrigation and absence of volumetric pricing are the possible reasons for this behaviour of the head end irrigators[7].

Changed Irrigation Requirement

The launching of CRDS, during the late 1950s, had a considerable impact on the cropping pattern in the command area. A large proportion of land was reported to have converted into paddy plots in line with the envisaged objective of the project. Sufficient supply of water facilitated three-crop paddy cultivation even in dry lands[8].

However, since late the 1970s, major changes occurred in the cropping pattern in the ayacut resulting in a significant decline in paddy and a sharp increase in the area under coconut, banana and plantain, arecanut and nutmeg. In effect mixed cropping replaced mono cropping of paddy in the ayacut. By mid 1980s rubber also acquired prominence among the crops. These shifts were more or less in line with the change in the cropping pattern, elsewhere in the State, which gained further momentum in 1990s.

In the field study, out of the 474 plots, only 175 had mono cropping (mostly paddy and rubber). The distribution of sample plots according to main crop cultivated (Table 4.1) captures the shift in cropping pattern in the area.

Table 4.1 : Distribution of Plots according to Main Crop Cultivated

Crops Grown	*No. of plots*	*Area (in cents)*
Paddy	124 (26.16)	22400 (39.15)
Coconut	251 (53.00)	24277 (43.08)
Banana and Plantain	20 (4.21)	1433 (2.50)
Nutmeg	1 (0.21)	151 (0.26)
Arecanut	1 (0.21)	209 (0.37)
Rubber	56 (11.81)	6989 (12.22)
Tapioca	2 (0.42)	118 (0.20)
Vegetables and other crops	15 (3.16)	1372 (2.40)
Fallow	4 (0.84)	263 (0.46)
Total	474	57212

Note : Figures in parentheses indicate the share to total.
Source : Survey Data.

The Table 4.1 shows that paddy is not the primary crop grown in the ayacut either in terms of number of plots or area-wise. Paddy claims prominence only in about one fourth of the plots, covering about 40 per cent of the area. Coconut occupies the top position in 53 per cent of plots and 43 per cent of area. Rubber, plantain and banana and vegetables are the other important crops grown respectively. The low proportion of plantain and banana, nutmeg, arecanut, and vegetables is because these crops are grown as mixed crops along with coconut. The subsidiary crop-wise distribution of plots justifies the importance of these crops in the study region (Table 4.2).

Table 4.2 : Distribution of Plots according to Subsidiary Crop Cultivated

Main Subsidiary Crops Grown	*No of Plots*	*Proportion of plots to total no. of mixed cropped plots*
Paddy	1	0.34
Coconut	54	18.31
Banana and Plantain	99	33.56
Nutmeg	77	26.10
Arecanut	36	12.20
Rubber	1	0.34
Vegetables and Others	27	9.15
Monocrop and Fallow	179	—
Total	474	—

Source : Survey Data.

The Table 4.2 shows that plantain and banana is the main subsidiary crop accounting for 99 out of the 295 mixed crop plots (33.6 per cent). The other important subsidiary crops are nutmeg (26.10 per cent), coconut (18.31 per cent) and arecanut (12.20 per cent). Thus, the situation in the command area is now mostly akin to that of mixed cropping. Coconut is the principal crop with plantain and banana, nutmeg and arecanut as subsidiary crops[9].

The shift in the cropping pattern in favour of non-paddy crops has resulted in significant changes in the demand for irrigation. Irrigation is now mainly required for coconut and mixed crops, the requirements of which differ from that of paddy. Non-rice crops, though demand less quantum of water compared to that of paddy, warrant more regulated water control to ensure timely and intermittent water depending on growth stages and the mix of crops grown. Further, for crops like nutmeg, coconut and arecanut failure in irrigation even for a short spell can lead to long term loss in production and productivity[10]. One important outcome of the changed cropping pattern, is the shift to rubber. This has resulted in a substantial increase of unirrigated plots in the command area. Out of the 474 plots surveyed, 77 (17.85 per cent) were unirrigated plots. Rubber accounts for about 77 per cent of these plots and most of these rubber holdings are situated in the tail ends.

Alternative Irrigation Institutions in CRDS

The overall picture that emerged at the time of survey is that of co-existence of multiple institutions in the provision of irrigation. To meet the challenges of uncertainty of water availability coupled with the changed crop requirements, farmers are found rely on alternative institutions in the canal command. Canal irrigation (public provision) is the common institution within which other institutions are evolved either as improvisations or alternatives.

These institutions could be perceived as rules, conventions, informal traditions and enforcement features concerned with the five irrigation management activities such as water acquisition, water distribution, maintenance, resource mobilisation and conflict resolution. These include rules for allocation and distribution of water, rules for mobilising resources for maintenance and other costs, courts and other State or local organisations responsible for resolving conflicts and enforcing rules for water distribution and resource mobilisation.

Accordingly, the irrigation institutions in CRDS command area identified during the survey can be broadly classified into six categories :

1. Public Provision,
2. Government Initiated Collective Action (in canal),
3. Farmer Initiated Collective Action (in canal),
4. Collective Action in Lift Irrigation,
5. Water Market,
6. Private Investment.

Table 4.3 outlines the characteristics of these irrigation institutions.

Table 4.3 : Characteristics of Irrigation Institutions

Institutions	*Characteristics*
Public Provision	Quantity and source of irrigation water, distribution, and maintenance and resource mobilisation of the system and conflict resolution are governed by the rules and norms of the State and farmers role is limited to water management below the outlet.
Government Initiated Collective Action	Quantity and source of irrigation water are governed by the rules and norms of the State. Rules and norms of distribution, maintenance, resource mobilisation and conflict resolution are largely framed by farmer groups within the guidelines stipulated by the State.
Farmer Initiated Collective Action	Quantity and source of irrigation water are governed by the rules and norms of the State. Rules and norms of distribution, maintenance, resource mobilisation and conflict resolution are mostly decided by farmer groups.
Collective Action in Lift Irrigation	Source of water is either public surface irrigation or a common pool resource (e.g. river). Quantity and source of irrigation water, distribution, maintenance, resource mobilisation and conflict resolution are decided by farmer or community groups within the overall socio-legal frame in the case of farmer/ community owned and controlled lift irrigation. In Government run lift irrigation schemes, these are governed by the rules and norms of the State.
Water Market	Source of water is either public surface irrigation or a common pool resource (e.g. river) on which right is created through location advantage. Water distribution, maintenance, resource mobilisation and conflict resolution are governed by the rules and norms set by the seller within the overall socio-legal frame.

(contd.)

(TABLE 4.3 contd.)

Institutions	*Characteristics*
Private Investment	Source of water is either public surface irrigation or a common pool resource (e.g. river) on which right is created through location advantage. Water distribution, maintenance, resource mobilisation and conflict resolution are governed by the rules and norms set by the individual farmer within the overall socio-legal frame.

Source : Survey Data.

There can also be cases where more than one of these institutions are jointly resorted to irrigate a single plot. In those cases, the one that is used most of the time is taken as the institution corresponding to those plots. The distribution of sample plots across various institutions is given in Table 4.4.

Table 4.4 : Distribution of Irrigated Plots and Area across Various Institutions

Irrigation Institutions	*No. of Plots*	*Irrigated Area*
Public Provision	58 (14.6)	6412 (13.6)
Govt. Initiated Collective Action	6 (1.5)	780 (1.7)
Farmer Initiated Collective Action	85 (21.4)	9398 (20.0)
Collective Action in Lift Irrigation	26 (6.6)	2557 (5.5)
Water Market	50 (12.6)	5924 (12.6)
Private Investment	172 (43.3)	21923 (46.6)
Total	397 (100.0)	46997 (100.0)

Note : Figures in parentheses indicate the share to total.
Source : Survey Data.

As many as 49 plots accounting for 37.5 per cent of the total plots depend on canal irrigation system (including collective action institutions). However, cases of sole dependence on the public provision system are low with 15 per cent both in terms of number of plots and area. Private investment in irrigation is found to cater to the largest proportion of plots in the sample (43.3 per cent). Area-wise also, it accounted for 46.6 per cent of the total. Water market assumes importance as it covers 12.6 per cent of the total area and plots. In brief, it can be inferred that alternative institutions assume significance in the command area of CRDS. These institutions are discussed subsequently along with case studies.

Government Initiated Collective Action

As discussed in Chapter II, involvement of farmers in the management of irrigation systems was the main strategy initiated by Government in the 1970s and 1980s. In conformity with this national agenda, Command Area Development Programme was implemented in the project area in 1978. Nevertheless, the programme became active only since mid 1980s with the passing of the Kerala CAD Act, 1986.

In CRDS, the CAD Programme envisages the promotion of participatory irrigation management under which the management of some of the systems is being transferred to the Water Users Associations. The programme is administered through an independent body within the Irrigation Department. CAD Programme is administered through a three-tier system consisting of a Project Committee, Canal Committee and Beneficiary Farmers Association (BFA). The organisational structures of these committees are similar throughout the State and have been explained in Chapter II. In terms of assumption of key roles, BFAs *(Karshaka Samitis)* are the grassroot bodies of the CAD Programme and are organised and registered for each outlet command, generally ranging up to 40 hectares. The primary activity of BFA is the construction and lining of field channels in 5-8 hectare blocks within the outlet command. They are also responsible for water distribution and maintenance below the outlet. Each BFA is supposed to design and implement a schedule for water distribution to farmers. Apart from these activities, the subsidised inputs are also channelled through BFAs.

The launching of the CAD Programme in its initial years in the CRDS command area induced some collective initiatives, which improved to some extent, the water availability and distributive efficiency of the canal. The number of *Karshaka Samitis* (BFAs) increased considerably during the first phase of CAD Programme[11] bringing in some changes in terms of construction, repair and lining of field channels and land levelling and shaping.

Most of the farmers surveyed in the areas, where CAD activities existed, have been part of it at some point or other. The subsidy component[12] was found to be the major attraction for the farmers in forming the BFAs. The importance of this institution in the past is evident from the fact that almost all the farmers in the CAD areas were members of the BFAs and had received some benefits.

Despite the above, the present position of the institution is dismal. During the survey, only 6 plots (1.5 per cent) were found, which were mostly depending on this institution. The poor working of the CAD system was pointed out as the major reason for moving away from this arrangement. The reasons cited for the unimpressive performance of CAD Programme are summarised as follows. Firstly, the working of the committees under CAD Programme is reported to be unimpressive. The ways in which BFAs are initiated and designed resemble those of a sponsored or directed body of government. The formal nature of the committees induced competition among the farmers for position as office bearers in the committee. Further, as the committees deal funds, there are complaints regarding the misappropriation of finance and undesirable nexus between CADA officials, engineers, contractors and the office bearers of BFAs.

Secondly, the CADA in its initial phases emphasised on field channel lining and rotation of turns, which increased the availability of irrigation water by minimising transmission loss. This also resulted in a more equitable distribution of water irrespective of the location of the plot. However, many farmers in the head-end opposed this, as it led to considerable reduction in the availability of water in their plots due to the arrest of seepage and enforcement of water turns. This in turn resulted in non-cooperation or passive representation by head-end farmers in BFAs leading to failure of CADA. Thirdly, scope for getting financial and input subsidies motivated many passive farmers to join BFAs, ultimately resulting in the poor functioning of the body. CADA could not also solve the problems related to the physical aspects of the project such as defective head regulator, damaged canal network and improper design. Finally, interference by influential farmers and politicians, bribing, biased and partial attitude of the irrigation officials and so on are reported as major hindrances towards the sustenance of the BFAs. Two case studies focusing on the functioning of the BFAs are given below.

CASE 4.I : PALAPPILLY BENEFICIARY FARMERS' ASSOCIATION

Palappilly Beneficiary Farmers' Association was formed in 1986 under CAD Programme. The formation of BFA, in fact, marked the formalisation of the activities underway on the already existing farmers' collective *Palappilly Karshaka Samiti*, which had a fairly impressive history. The organisation, initiated in the 1970s, was as an outcome of the growing scarcity of water in the region. The unreliability and inadequacy of canal water in irrigating the crops was the major reason, which induced collective action. In the initial years of the CRDS, paddy was grown in all plots and in many a cases, with three crops. Assured irrigation resulted in conversion of even garden lands to paddy fields. Neighbouring localities of Palappilly also experienced similar patterns of land use and intensified paddy cultivation. These resulted in a growing demand for canal water, which culminated in acute scarcity by the mid 1970s. A crop failure in 1977 gave the immediate inspiration for cooperation among farmers. A group of about 20 farmers in the region under the leadership of a socially accepted and politically powerful big farmer carried out the maintenance of canal, distributaries and field channels through locally mobilised finance. These efforts resulted in a commendable improvement in the water supply.

The CADA officials, looking for potential farmer communities for initiating such group action, provided a formal structure to the organisation as Palappilly BFA (popularly called Palappilly CADA Committee). The formal structure and the subsidies associated with the Committee attracted an increasing number of farmers. The BFA has a membership of about 60 farmers with a total area of around 100 acres. Out of these, 6 were big farmers with more than 3 acres of land. In its initial years of formation, the Committee constructed and lined the field channels, organised soil testing and agricultural extension services and channelised subsidies. Within 5 years about 60 per cent of the field channels were lined. The BFA also designed and implemented a rotational system with specific time spells for each member, according to the size of plot.

Towards the mid 1990s, the functioning of the BFA started facing problems, which became severe by late 1990s. The reasons for the failure of the BFA were manifold. At the outset, the formalisation of group action through BFA, converted the leaders of the organisation into 'office bearers'. The official powers assigned to the leaders resulted in unhealthy competition among members to assume important positions in the organisation and those who did not succeed, started opting out of the activities of the BFA. The fact that CADA committees managed the funds allotted by the Government for land upgradation exposed the selfish motives of certain members, which provoked allegations and counter allegations amongst them.

In the meanwhile, increased wages of agricultural labour, declining profitability of paddy, aversion of younger generation towards farm work and the growing tendency of part-time cultivation and so on resulted in changing the crop pattern in favour of non-paddy crops and to leasing out substantial land on rent. Those who switched over to non-paddy crops in the initial years were mostly big farmers which in turn resulted in the decline of BFA as a prominent pressure group in the area.

Further, as the activities of the BFA included lining of field channels and introduction of strict monitoring of water-turns, the farmers in the head-ends began to oppose the activities of BFA as it affected the availability of seepage water to their plots. The head-end users who had access to water in abundance are dissatisfied as the water is now shared more equitably among all farmers irrespective of location. This has led to the destruction of CADA field channels, leading to many incidents of canal destruction, which resulted in litigation, in few cases. The members also complained of violation of schedules and misappropriation of water by few influential farmers. All these led to non-cooperation of some farmers. The absence of able leadership subsequent to the withdrawal of the earlier leaders also affected the organisation.

The changed priorities of the Government vis-a-vis the CAD Programme reflected on the activities of the BFA, which were met through the contributions from the members before the formation of BFA. This system vanished with the flow of funds from CAD Programme, which in later years became inadequate and untimely for meeting the maintenance cost of the system. As a cumulative effect of these inadequacies and limitations, the Palappilly BFA has become redundant in terms of activities, though it continues to be a registered association. Most of the farmers, even the members of the BFA have now invested on wells and pump sets in their plots.

CASE 4.II : PUTHENKARA BENEFICIARY FARMERS' ASSOCIATION

Founded in 1987, Puthenkara BFA is the outcome of a collective action initiated by CADA under Thanguchira Branch Canal area. Spurts of collective action were present in the locality ever since the branch canal was in function. However, these were mostly spontaneous and short-lived, such as spout destruction and collective maintenance of distributaries during seasons of acute shortage. The farmers are mostly paddy cultivators and more than 80 per cent of the land was under paddy.

The formation of BFA, under the CAD Programme, gave a fillip to the collective activities. Presently, there are about 30 farmers in the BFA with around 45 acres, of which a considerable proportion is under paddy. The other crops are mainly coconut followed by nutmeg and arecanut.

The BFA, by putting pressure in the project committee succeeded in enforcing a time schedule for supply of water to the region. Accordingly, water is available in the canal during 10-12 days in a month, with 2-day spells at an interval of 5 days. As per the existing rule, water is supplied for 30 minutes per acre of land. The farmers strictly adhere to this schedule and hence there are no conflicts among them.

Apart from the involvement in the distribution of water, the BFA also takes up the contract for maintenance of the branch canal (above the outlet) with the co-operation of farmers in the neighbouring villages. Maintenance and construction of field channels, agricultural extension services, soil testing etc. are other tasks carried out under the BFA. The BFA has been acclaimed as a model of State-nurtured collective action in the left bank canal.

Farmer Initiated Collective Action

Farmer initiated collective action refers to joint action initiated by farmers[13] to improve water availability in terms of quantity and supply at regular intervals within the canal system. The activities include group participation in the maintenance of canals, designing, scheduling and enforcement of water turns, engaging canal watchmen and renovation and construction of reservoirs. Joint action could also take the form of organising democratic protests (*dharna*, road picketing), petitioning and submission of memorandums, exertion of political influence, bribing of officials and other staff, spout destruction and so on, which are intermittent or once for all. While democratic protests and petitioning are much within the rights of individual farmers, other forms do raise issues of corruption and equity. The second set of actions is mostly spontaneous in nature with short term impacts and do not have any defined rules and hence do not fall under the institution of farmer initiated collective action[14].

The study defines farmer initiated collective action as activities taken up by a group of farmers to improve availability of and equity within the distribution of water among users, which assumes the form of a formal and continued institution with long term implications. Farmer initiated collective action has a long history in the command area and the farmers recollect their concerted efforts in improving water flows even during the initial years of CRDS. A major organisation which strengthened farmers collectives of late is the group farm committees (Padasekhara Committees) introduced by the Agricultural Department during late 1980s to practice group management in paddy cultivation[15]. Most of the farmers' collectives assumed a definite form with the initiation of group farming.

At the time of survey, farmer initiated collective action is found as a major institution in the canal command constituting about 21 per cent of plots and 20 per cent of area. Various forms of collective actions were observed across 85 out of 397 surveyed plots, which are given in Table 4.5.

Table 4.5 : Distribution of Plots across Various Forms of Collective Action

Types of Collective Action	*No. of Plots*	*Percentage to Total*
Joint action for Group Farming	46	54.12
Canal maintenance, construction of field channels etc.	17	20.00
Design and implementation of irrigation turns	15	17.65
Renovation and digging of reservoirs	4	4.71
Overall activities	4	4.71
Total	85	100.00

Source : Survey Data.

The major form of farmer initiated collective action is the action taken under the group-farming committees (54.12 per cent of plots). The formation of such committees is facilitated by the input and subsidy incentives[16]. The other important forms of farmer cooperation were visible in maintenance of canal, construction of field channels and in designing and implementation of irrigation schedules.

Despite their prominence, farmers' initiatives are on the decline as per the respondents of the survey. Uncertainty in the water availability due to perceived deficiencies of the canal system and the inequities in sharing benefits of cooperation are the prime reasons indicated as the refraining factors. Two case studies on farmer initiated collective action are given below.

CASE 4.III : ASHTAMICHIRA—THAZHEKKAD FARMERS

The problem of water scarcity had been an acute form in Ashtamichira Thazhekkad Region till mid 1980s. Though the area was under the ayacut of CRDS, the distance from the spout and its position was unfavourable to the farmers. The resultant scarcity of water, induced cooperation among the farmers. In the first instance, collective action was in the form of spout destruction and seizing of water turns by bribing the lascars. However, these actions were found unsustainable solutions.

The extension of the canal by putting pressure on irrigation authorities was the next step followed by the farmers. Towards this, a Committee was formed with about 30 members, with medium sized plots (50-200 cents). Paddy was the main crop in these plots followed by mixed crop coconut. The pressures exerted on the authorities took the form of submitting mass memoranda followed up with protests and *dharanas* and exertion of political pressure. The presence of the Panchayat President and few socially accepted and powerful persons (school teachers, civil servants) among the farmers was the added advantage to the Committee.

These concerted activities resulted in an extension of the canal, *Ashtamichira-Thazhekkad Canal Extension*, of about 400 metres. Following this, the committee also renovated a huge Panchayat tank into which the water is diverted and stored. The collected water is distributed among the farmers with definite turns, the timing of which is decided on group consensus. Each plot is supplied with water for 30 minutes per acre, continuously for 6 days, with an interval of 6 days. In order to maintain the turn and operate the shutters, one person is engaged from within the members. To meet the running expense, including the honorarium to the operator, each member pays an amount of Rs. 30 per acre per month. In addition to this, small contributions are collected from the members from time to time towards meeting the other expenses. The average annual expenses are about Rs. 100 per farmer. The farmers are satisfied with the Committee, which has completed more than one decade of its activities.

CASE 4.IV : FARMERS' ACTION IN ADICHILI BRANCH CANAL

Adichili is the first branch of the Left Bank Canal and has the first spout of LBMC. Though it has the advantage of head-end, due to the fixing of the spout at a higher plain, scarcity of water is felt during peak seasons in the middle end and tail ends of the branch. The crop pattern of the farmers is predominantly mixed with nutmeg and plantain followed by coconut and arecanut. At times of water scarcity, farmers cooperate and collectively put in efforts to divert the water by blocking the flow in the main canal (this is locally referred to as *kedayidal* blocking). This is a recurring activity in every season marked by water shortages and has the support of all farmers as it brings immediate benefits. This apart, every year, the farmers collectively maintain the canals and field channels.

Collective Action in Lift Irrigation

Collective action in lift irrigation refers to the small scale irrigation systems designed for the distribution of water from sources such as rivers, ponds, wells, other natural water bodies and artificially constructed reservoirs. The major form of collective action in lift irrigation in CRDS command area is that where water is lifted from sources and distributed to the members of the community. 26 plots were identified in the survey depending on collective action in lift irrigation for water. The details of these schemes are summarised in the following Table 4.6.

Table 4.6 : Details of Collective Action in Lift Irrigation

Pump Capacity	*No. of Plots*	*Source of Water*		
		River	*Ponds/ Wells*	*Artificially* Constructed Reservoirs
10-20 HP	11	1	7	3
21-40 HP	2	1	—	1
41-80 HP	13	13		—
Total	26	15	7	4

Source : Survey Data.

The Chalakkudy River is found as the major source of collective action in lift irrigation. The other two sources are mainly dependent on canal water—reservoirs are constructed at locations of canal water drainage. Collective action in lift irrigation in the CRDS command can be divided into 4 types, according to ownership and management : (a) owned and operated by Government with farmer participation (4 cases); (b) Government owned and farmer managed (3 cases); (c) owned and managed by farmer groups (12 cases); and (d) run by community and social groups (7 cases). A large proportion (about 70 per cent) of the farmers under collective action in lift irrigation are now exempted from paying cess. The exemption was obtained after years of litigation following various forms of pressure put on various government authorities.

The number of plots that are under collective action in lift irrigation ranges from 15-200. Despite variations across projects, fairly strict schedules are maintained through the engagement of pump operators and water regulators. The operation costs of the projects are normally recovered through water charges. Water prices are charged either on monthly basis per acre or time basis (per hour/per acre). Generally, these were charged on time basis in projects with comparative water scarcity. Per hour charges range between Rs. 25-35 (per acre) while monthly charges vary between Rs. 50-100 per acre per month. Irrespective of the ownership and management of collective action in lift irrigation, government subsidies on purchase of pumpsets, construction of reservoirs and subsidised power charges facilitated the emergence of the institution.

CASE 4.V : POOLANI LIFT IRRIGATION SCHEME

Poolani-Chettithope region comes under the ayacut of Poolani Branch Canal of the CRDS Left Bank Canal. Despite the existence of State regulations restricting the expansion of lift irrigation, in the ayacut of an already existing irrigation scheme, the farmers of the region got organised with the objective of setting up of a lift irrigation scheme. Except during the few initial years, the CRDS canal failed to provide water to the region leading to severe scarcity. There were also instances of paddy failure during these periods.

This uncertainty and inadequacy in the availability of water from the canal and possibilities of farming high-value crops such as coconut, arecanut and nutmeg (due to their favourable prices compared to paddy) prompted the farmers of *Poolani Lift Irrigation Samiti* to go for alternative institutions. Initially the farmers tried to introduce a lift irrigation scheme with 50 per cent financial assistance from Small Farmers Development Association (SFDA), a developmental organisation of the State. However, the idea was dropped when the possibility of a fully government sponsored lift irrigation scheme came into consideration. The *Samiti* pressurised the authorities and was successful in getting an exemption for the farmers under it from the ayacut of the major scheme and convincing the authorities the scope for a lift irrigation scheme due to their access to Chalakkudy River.

Accordingly, in 1984, the Poolani Lift Irrigation Scheme started with an envisaged ayacut of 135 acres. The scheme is owned and operated by minor irrigation wing of irrigation department. Pumping is done by a motor of 90 HP, run by electricity, with a stand-by motor of same capacity. The allocation of water is done on turn basis with the help of pump and spout operators. The distribution of water starts from the tail end of the channel in order to ensure water availability to all the members. The cess is collected through village office (at the rate of Rs. 63 per acre per year).

A total of about 120 members, all farmers initially within the ayacut of CRDS, are benefited by the scheme. Among these 25 are small farmers with less than 20 cents of land. More than 90 farmers have holding size of 2–5 acres and the rest are with 5–8 acres. Almost the entire area is now converted to mixed crop with coconut, nutmeg, plantain and arecanut. Vegetable cultivation is also practiced in some plots, seasonally.

Even after the commencement of the lift irrigation scheme, water scarcity was not resolved affecting the viability and prospects of garden crops like nutmeg and plantain. This was due to the inefficiency in running the system. This led the farmers to take over the management of the scheme. Three watchmen were appointed for running the system, the salaries of whom are borne by the members for which a monthly collection of Rs. 40 per acre. Apart from this, the Committee, through locally mobilised funds, carries out small maintenance works.

CASE 4.VI : MANGALATHUTHAZHAM LIFT IRRIGATION PROJECT

The Mangalathuthazham Lift Irrigation project is a lift irrigation scheme launched under the aegis of the Church in Karukutty Kariamparambu in 1993. The farmers in this area are mostly small holders and had water problems since the branch is located towards the tail end. Due to water scarcity most of the plot cultivated tapioca or other dry crops. As most of these farmers belong to Catholic community, the parish church took initiative in organising the farmers to set up a lift irrigation scheme, using the water from the nearby Mangalathuthazham River. The farmers under the church collectively went for litigation to get exemption from the CRDS ayacut. Some farmers got exemption on account of high lying plots in 1991, while the others are still within the ayacut and pay cess.

80 HP pump set and a pump station have been set up with a cost of roughly Rs 5 lakh and the major portion of the cost was borne by the church. Apart from this a subsidy from agricultural department and some contributions from few big farmers were also used to meet the initial cost. A committee of beneficiaries is formed, which under the guidance of the church fixes and implements the schedule and collects the water rates. The rate for water is fixed at Rs. 24 per hour per acre. The coming of the scheme led to the change in the crop pattern of the area towards more of mixed cropping with nutmeg and coconut as the main crops.

Water Markets

The term 'water market' refers here to localised, informal arrangements through which owners of water-abundant plots sell water to other farmers in the community. In the command area of CRDS, three types of water transactions are seen : (1) Sale of canal water turns; (2) Sale of well/pond water dependant on canal water; and (3) Sale of water from other sources such as river.

Of these, sale of water turns is rare but is significant among those farmers who have shifted towards less water intensive crops such as rubber. These farmers are mostly at a disadvantageous position in respect of water availability and hence, water turns are sold temporarily to farmers at the head end of the water-course. During the field survey, few cases were found conforming to this pattern. The water charges were found more or less equivalent to the water cess to be paid to the authorities in most of the cases. However, in certain cases, the charge was found higher than the cess. The higher charge was attributable to the higher availability of water and the profitability of the crop for which irrigation water is provided.

The sale of well water, dependant on canal water, is the most common form of water markets in the area. Farmers located in favourable positions, in terms of water availability, were found investing on wells and pump-sets so that water obtained from these could be sold to neigbouring farmers, who do not receive adequate and reliable canal water. Water is lifted from wells constructed close to the canal. Most of the respondents of the survey admitted that though it appears as a private resource, the main source of water is the seepage from the canals[17]. Water is distributed to buyers either through unlined field channels or through lined channels[18] or PVC pipes.

The buyers are generally of two types : those who depend on water market only during periods of acute water scarcity *vis-a-vis* the crop requirements; and those who rely solely on water market for the entire season[19]. The sellers are generally big farmers or those who are economically well off and thus capable of investing an initial sum for instituting the system. In many cases, they are principally farmers themselves, who supply the excess water in their wells to neighbours. The main attraction of investing in the system is the possibility of utilizing a significant proportion of its capacity in own holdings. The fixed investment under the institution is generally borne by the seller and water charges are fixed on an hour basis according to the power of the pump. As has been noted by Shah, once such investments are made, rational behaviour precludes the consideration of fixed costs in their pricing of water and they are guided mainly by the marginal cost of extracting and delivering water (Shah, 1985). In cases, where the sellers are primarily water traders, the fixed cost considerations are also entered in the pricing.

The third category refers to lifting of water from the river or ponds, which do not have any direct linkage with the canal system. The sellers, as in the case of well-based water markets, are mostly farmers themselves who cater to the requirements of 4-5 neighbours with the excess capacity of their pump-sets. The pricing considerations are also similar to that of well-based system.

Primary attention in the study was given to the second and third categories as they assume significance in Chalakkudy Command and are comparable. The survey found 50 cases of water-purchase of which 41 belong to well-based water market and the rest were based on river water. The following Table 4.7 gives the details of distribution of plots under water markets.

Table 4.7 : Details of Water Markets

Pump Capacity	*No. of Plots*	*Fuel Type*		*Pumping Hours per Acre*	*Water Price per Hour (Rs)*
		Electricity	*Diesel*		
Upto 5 HP	17	14	3	3-5 hours	8-11
6-10 HP	29	20	9	2-3 hours	10-15
11-15 HP	3	-	3	1-2 hours	12-20
More than 15 HP	1	-	1	< 1 hour	15
Total	50	34	16	-	11.91*

Note : * Indicates crude total average as it is calculated from ranges.
Source : Survey Data.

34 per cent of the water sellers use pump-sets with capacity upto 5HP. A large proportion of these sellers are found to have purchased the pump-sets availing subsidy from Agricultural Department. It is observed that 6-10 HP pump-sets are commonly used by water sellers. 58 per cent of the sellers belonged to this category. The major source of power used in pumping water is electricity (68 per cent). However, in large pump-sets, the fuel used is diesel. The increased preference for electric pump-sets is attributed to the availability of power at subsidised rates for agricultural purpose. The water prices (per hour, per acre) varied in the range of Rs 8-20 depending on the types of fuel and the capacity of the pump-sets. The price of water pumped using diesel pumps was found Rs. 2-4 dearer compared to that of electricity based systems.

CASE 4.VII : SALE OF WATER FROM POND IN MANJAPRA

In Manjapra, water is sold by a farmer to 8 neighbouring farmers engaged in dry land farming. The source of water is a pond dug in the seller's paddy field. The location of the plot favours storing the drainage water from the canal, which is the source of water in the pond. The plots of all the buyers are located above the normal ground level. CAD Committee exists in the area and most of the buyers were earlier members of the Committee.

Water is lifted with the help of using an 8 HP motor and supplied through channels constructed as a part of the CAD Programme. The pump was purchased at subsidized rates—the subsidy being provided by the Agricultural Department (20 per cent). The buyers and the seller come within the ayacut of the CRDS. Water is bought mainly for cultivating banana, nutmeg and coconut.

As the seller is a farmer and requires water in his own plot, he undertakes to sell the water in his spare time. Some of the buyers buy water only to supplement the canal water and water from their own wells, while others are fully dependent on the seller for their irrigation requirement. The water rates are fixed on an hourly basis at the rate of Rs. 12 per hour/per acre. The irrigation turns are fixed informally on mutual consensus with flexible rotational practices. The predominance of water sensitive crop like nutmeg and banana are the main reasons for ensuring certainty in the supply of water.

CASE 4.VIII : SALE OF WATER FROM RIVER IN MELUR

A system of lifting and sale of river water by the farmers who have plots adjacent to the Chalakkudy river has been a common practice in Melur Panchayat for the past 10-15 years. About 20 such water markets exist in the region catering to more than 200 plots located in disadvantageous positions. In one such case, the seller has installed a 15 HP pump-set and lifts water from the river. The seller is not a full time farmer, though has a small plot of banana cultivation. Water is sold to 18 farmers in the nearby plots through earthen channels constructed by the buyers. All the buyers' plots come under the ayacut of CRDS and were involved in a multitude of collective efforts to improve the supply of water prior to their choice of water purchase.

Water is bought mainly for plantain[20] cultivation. The irrigation requirement of plantain is unique as it requires more regulated and timely water supply. The farmers' time calculation of harvest also largely depends on the scheming of irrigation. For plantain plots the water charge is Rs. 8 per plant for the season, and an unlimited quantity is supplied according to the need of the buyer. For other mixed crops, the charges are Rs. 12 per hour/per acre. No formal turn basis is observed and water is pumped according to the requirements of the buyers.

CASE 4.IX : SALE OF WATER FROM WELL IN KALADY

In the lower reaches of Kalady Branch Canal, several small scale water markets exist. In most of these cases, the seller caters to one or two farmers in the neighbouring plots. The water is pumped from the well in the sellers' plot, which is located advantageously to store canal water. Water is bought to supplement the erratic supply from canal in times of scarcity. The sellers themselves are farmers with small plots and sell their excess water to the buyers.

In many cases PVC pipes are installed to transfer water, the cost of which are borne by the buyers. CAD field channels and earthen channels are also used for this purpose. The pump sets are mostly of 3 HP bought with the help of government subsidy. The pricing is often done in such a way as to recover the running cost from the buyers and the unit price ranged between Rs. 6–8 per acre/per hour.

Private Investment

Private investment in irrigation in the command area is in the form of construction of open wells and setting up of pump sets on an individual basis. Wells account for the largest share of irrigated area among the surveyed plots (about 46 per cent of total area)—172 out of 397 plots studied were with at least one well out of which 55 had two wells (Table 4.8).

Table 4.8 : Distribution of Plots under Wells

Pump Capacity	*No. of plots with one well*	*No. of plots with second well*	*Total number of wells*
1.5 HP	41	30	71
2 HP	5	1	6
3 HP	115	23	138
4.5 HP	—	—	—
5-10 HP	11	1	12
Total	172	55	227

Source : Survey Data.

The main source of water in the private wells is the seepage from the canal system. This is evident from the fact that water levels of the wells go down substantially when there is no water flow in the canals. Farmers perceive wells as a means to store water, which allows enhanced flexibility and certainty in irrigation. Most of the farmers who invested on wells and pump-sets were found to have availed the subsidies from the Agricultural Department. In 165 cases (about 73 per cent) farmers had availed subsidy. Around 97 per cent of the farmers received more than 40 per cent of the installation cost as subsidies. Furthermore, the subsidy on electricity for agriculture use has facilitated private investment.

CASE 4.X : OPEN WELLS IN PARAKKADAVU

In Parakkadavu, a farmer located very near the main canal has two open wells with 3 HP and 1.5 HP motors for irrigating the plot of 150 cents. Mixed cropping is practiced with coconut as the main crop and nutmeg as the inter crop. Being at the head-end, till the coming of the canal committee in 1992, the plot never had water problems. However, with the strict scheduling of irrigation turns introduced by the local canal committee, the water availability was restricted to certain days and timings—at times, water was available only during nights or at odd times. The plot had the locational advantage, enabling the availability of seepage and hence investment on wells was found to be more convenient. Further, the part-time cultivator status of the farmer provided the motivation for digging well to reduce time spent on irrigation. The well was constructed and a 3HP motor pump was installed using the subsidy (20 per cent) from Agriculture Department. A second pump was also installed in the well used for domestic purpose, at own cost.

CASE 4.XI : INVESTMENT ON PUMP-SET IN KARAYAMPARAMBU

In Karayamparambu, a farmer in the tail end has undertaken investment on well and pump-set. The plot (60 cents) comes under the original command area of the CRDS scheme but was exempted from water cess in 1989. Barring few garden trees the plot was non-cultivated. It was included in the ayacut of a newly started lift irrigation scheme, Mangalathuthazham Lift Irrigation (formed and run under the aegis of the local church). After getting associated with the lift irrigation scheme, water availability was fair and the plot was converted for plantain and coconut cultivation. The monthly water charge was Rs. 42. With the functioning of the lift irrigation, the water level of the well in the plot improved considerably due to seepage. The availability of adequate amount of water in his own plot prompted the farmer to leave the membership of the lift irrigation scheme. Subsequently, a pump-set was purchased with the help of a loan from the Bank and partly with the subsidy from Agriculture Department.

From the above discussion, it is clear that the failure of the surface irrigation system to meet the requirement of farmers has led to the emergence of multiple institutions. However, the choice of institutions is found to differ across farmers and plots. Detailed examination of these factors is attempted in the subsequent chapter. There are a number of factors such as policies of the Government, prices of commodities and costs of inputs, availability of subsidies and so on which cannot be considered in explaining the presence or absence of an institution in a particular locality since such factors do not vary much across farmers involved in different institutions. Thus, it could be hypothesised that a farmer's selection of a particular institution is determined by the following : locational, land and crop, and individual and household characteristics. This is verified in the subsequent chapter.

NOTES

1. In certain portions in the RBMC the bed width of take is 7 m. Depth of water is 2 m; slope, cutting 1 :1 and filling 11/2 :1; gradient 1 in 1760 at head with a velocity of 0.97 m/sec at head.
2. The figure shows the lay-out of a typical canal network, where the arrows show the direction of the water flow. Water from the river is fed, through a regulatory mechanism (dam/weir/barrage), into the main canal which splits into branch canals which further branches out into distributaries and finally into minors. Outlets are provided either on the distributaries or on the minor canals. Water-course starts from the outlet, which further splits into field channels. The water availability varies considerably depending on the position of the plot. This warrants the recognition of the importance of locational factors in any study on canal-based irrigation systems.
3. During the survey, a new branch canal (Mundapallypadam Branch Canal) of 1 kilometre length was noticed, which do not confirm with the official map obtained from the Irrigation Department. From further enquiry, it was understood that the same was catering to the fields near the residence of the sitting Member of Legislative Assembly.
4. Whether this decline in corruption is due to loss of confidence of farmers in lascars and other officials in a growing uncertainty regime or is it on account of other changes (crop changes, alternative possibilities for irrigation) is an issue, discussed in a later part.
5. Farmers near Mambra Branch Canal recollect the days when water was taken to the areas coming under the constituency of a very powerful political leader with the protection of police and irrigation officials.
6. However, there exist no unanimity, even among irrigation officials in this regard. For instance, the project report prepared by the irrigation department says that CRDS receives more than required water from the reservoir (Government of Kerala, 1994).
7. It is interesting to note that some of the irrigators at the head-ends reported of 'water scarcity' as they could not flood the farms as frequently as before.

8. A quantitative analysis of the effects of CRDS carried out by the Planning Board in 1967 observes that the project has resulted in a 50 percentage increase in gross area under paddy; 57.36 per cent increase in production and 819 pounds increase in productivity per acre. 55 per cent single crop area was converted into double crop and 2.5 per cent to triple crop. (*Evaluation of Some Minor Irrigation Projects in Kerala,* Bureau of Economics and Statistics, Government of Kerala, 1967).
9. The shift in cropping pattern was mainly due to profitability calculations of the farmers. The higher cost of cultivation coupled with the low price of paddy vis-a-vis other crops has been found as the crucial factor (Unni, 1983; Jose, 1991).
10. During the survey many farmers complained of the damage of crops, especially nutmeg due to untimely availability of water.
11. Number of BFAs under CRDS : Phase I

Year	*No. of BFAs*
1986-87	93
1987-88	297
1988-89	37
1989-90	17
1990-91	22
Total	466

Source : Irrigation Department, Government of Kerala.

12. Subsidies are given to farmers to cover expenditures on water distribution to the individual holdings with the WUA area. Each WUAs are given Rs. 100 per hectare for the first year and the second year and Rs. 75 per hectare for the third year not exceeding 4000 to each WUA.
13. This is different from government initiated collective action since the initiatives are taken by the farmers. They carried out either the same activities as that of CADA without its support or those activities which are beyond the purview of CADA
14. Some of these actions are found to be resorted to by groups engaged in long term cooperation as strategies.
15. The programme which was introduced in 1989 aims at increasing the viability of paddy through enhanced production and productivity, reduced cost of cultivation *vide* group action among paddy farmers.
16. As in the case of CADA, in group farming, subsidies and other inputs are channelised through farmers organisations—Padasekhara Committees.

17. At times water is lifted directly from the canal and sold to the farmers. One such instance was noticed in Edakunni village, where a seller provides water to about 6 fields (about 500 cents) using a 10 HP diesel pump.
18. Field channels constructed by CADA are used for this purpose in many cases.
19. The buyer in both these cases are also to bear the water cess charged by the authorities as they are coming under the canal command.
20. Melur is well known for banana and plantain cultivation.

5

Determinants of Institutional Choice in Irrigation

As seen in the earlier chapter, multiple institutions exist in the distribution of irrigation in the command area of CRDS and the choice of institutions is found differing across farmers and plots. It is reasonable to expect that farmers choose a particular institution from the set of alternatives available, comparing the expected pay-offs from various institutions. The costs-benefits calculations and hence the matrix of pay-offs of institutions differ across plots and individuals, which in turn depend on a host of factors. The present chapter discusses these determinants of institutional choice in irrigation.

Determinants of Institutional Choice in Irrigation

From the literature and the case studies discussed in previous chapters, three sets of factors can be delineated as the major determinants in irrigation decisions of the farmers. These are : (a) locational factors; (b) land and crop specifications; and (c) individual and household characteristics. The nature and extent of the relationship between these factors and their influence in the institutional choice are analysed in this section.

Locational Factors

An important determinant in the choice of the farmers is the location of the plot. The distance from the main canal, branch canal, the outlet, and elevation of the plot are the parameters taken to explain the location specific characteristics.

(a) Distance from the Main Canal : The number of plots depending entirely on canal/public provision decreases as distance from the canal increases. This is evident from the fact that out of the 58 plots depending on canal for irrigation 46 (79.31 per cent) which are within 16 km from the main canal (Table 5.1). Institutions of collective action (both government induced and farmer initiated) are also concentrated within this range. Well and water markets show more concentration in the middle reaches. All the plots under collective action in lift irrigation are beyond 12 km from the main canal. Thus, it can be inferred that distance from the main canal has some influence in the institutional choice of farmers.

Table 5.1 : Distribution of Institutions according to Distance from the Main Canal

Institution	*Distance from the Main Canal (in km)*						
	Upto 10	10-12	12-14	14-16	16-20	20 and above	Total
Public Provision	6 (10.34)	17 (29.31)	12 (20.69)	11 (18.97)	6 (10.34)	6 (10.34)	58 (100.00)
Government Initiated Collective Action	—	5 (83.33)	—	1 (16.67)	—	—	6 (100.00)
Farmer Initiated Collective Action	11 (12.94)	21 (24.71)	18 (21.18)	14 (16.47)	8 (9.41)	13 (15.23)	85 (100.00)
Collective Action in Lift Irrigation	—	—	4 (15.38)	7 (26.92)	7 (26.92)	8 (30.77)	26 (100.00)
Water Market	—	9 (18.00)	21 (42.00)	18 (36.00)	2 (4.00)	—	50 (100.00)
Private Investment	20 (11.63)	33 (19.19)	36 (20.93)	33 (19.19)	34 (19.77)	16 (9.30)	172 (100.00)
Total	37 (9.31).	85 (21.41)	91 (22.92)	84 (21.16)	57 (14.36)	43 (10.83)	397 (100.00)

Note : Figures in brackets are percentages to total.
Source : Survey Data.

(b) Distance from the Branch Canal : Distance from the branch canal is found to influence the choice of institutions that the farmer depends. Thus, the data shows definite pattern in the distribution of institutions : The number of plots depending entirely on canal irrigation is high among those located near the branch canal (Table 5.2), which could be attributed to the better availability of canal water in plots closer to branches. Farmer initiated collective action (76.47 per cent) and private investment (69. 77 per cent) is highly concentrated within 2.5 km. Collective action in lift irrigation is seen only in plots beyond 2 km from the branch canal. Government initiated collective action also show a concentration around the mid reaches. Thus, it is reasonable to expect a correlation between institutional choice and distance from the branch canal.

(c) Distance from the Outlet : Distance of the plot from the outlet is also presumed to have some influence in determining the institutional option for irrigation. The data given in Table 5.3 provides certain visible trends. The Table 5.3 shows that canal irrigation, collective actions and private investments are higher in the plots nearer to the outlets. 63.79 per cent of plots under public provision were found within 0.25 km. from the outlet. In the case of Government initiated collective action, farmer initiated collective action and private investment, the percentages were 50.00, 57.65 and 69.19, respectively. This could be because of the fact that all these institutions primarily rely on canal water (directly or indirectly). As expected, water markets[1] and collective action in lift irrigation is found concentrated mostly in plots distant from the outlets.

(d) Elevation of the Plot : Apart from distance from canal network, elevation of the plot is also recognised as an important factor in determining the choice of irrigation institutions. The data furnished below (Table 5.4) also corroborates this understanding.

Table 5.2 : Distribution of Institutions according to Distance from the Branch Canal

Institution	*Distance from the Branch Canal (In km)*									
	0-0.5	0.5-1	1-1.5	1.5-2	2-2.5	2.5-3	3-3.5	3.5-4	4 and above	Total
Public Provision	8	9	7	6	8	9	5	3	3	58
	(13.79)	(15.52)	(12.07)	(10.34)	(13.79)	(15.52)	(8.62)	(5.17)	(5.17)	(100.00)
Government Initiated	—	1	—	1	2	2	—	—	—	6
Collective Action		(16.67)		(16.67)	(33.33)	(33.33)				(100.00)
Farmer Initiated	6	18	20	12	9	6	5	5	4	85
Collective Action	(7.06)	(21.18)	(23.53)	(14.12)	(10.59)	(7.06)	(5.88)	(5.88)	(4.71)	(100.00)
Collective Action in	—	—	—	—	3	6	8	5	4	26
Lift Irrigation					(11.54)	(23.08)	(30.76)	(19.23)	(15.38)	(100.00)
Water Market	—	1	6	17	11	9	6	—	—	50
		(2.00)	(12.00)	(34.00)	(22.00)	(18.00)	(12.00)			(100.00)
Private Investment	18	36	26	18	22	19	16	11	6	172
	(10.47)	(20.93)	(15.12)	(10.47)	(12.79)	(11.05)	(9.30)	(6.34)	(3.49)	(100.00)
Total	32	65	59	54	55	51	40	24	17	397
	(8.06)	(16.37)	(14.86)	(13.60)	(13.85)	(12.85)	(10.08)	(6.05)	(4.28)	(100.00)

Note : Figures in brackets are percentages to total.
Source : Survey Data.

Table 5.3 : Distribution of Institutions according to Distance from the Outlet

Institution	*Distance from the Outlet (In km)*						
	0-0.1	0.1-0.25	0.25-0.50	0.50-0.75	0.75-1	1 and above	Total
Public Provision	22 (37.93)	15 (25.86)	9 (15.52)	5 (8.62)	6 (10.34)	1 (1.72)	58 (100.00)
Government Initiated Collective Action	1 (16.67)	2 (33.33)	—	2 (33.33)	1 (16.67)	—	6 (100.00)
Farmer Initiated Collective Action	23 (27.06)	26 (30.59)	22 (25.88)	7 (8.23)	7 (8.23)	—	85 (100.00)
Collective Action in Lift Irrigation	—	5 (19.23)	7 (26.92)	8 (30.77)	3 (15.54)	3 (11.55)	26 (100.00)
Water Market	5 (10.00)	5 (10.00)	18 (36.00)	12 (24.00)	9 (18.00)	1 (2.00)	50 (100.00)
Private Investment	41 (23.84)	78 (45.35)	36 (20.93)	8 (4.65)	6 (3.49)	3 (1.74)	172 (100.00)
Total	92 (23.17)	131 (33.00)	92 (23.17)	42 (10.58)	32 (8.06)	8 (2.02)	397 (100.00)

Note : Figures in brackets are percentages to total.
Source : Survey Data.

Canal irrigation and collective action are the major types of institutions in those plots which are located at a lower plain *vis-a-vis* the outlet. As against this, private investment and water markets are concentrated in plots, which lie above the outlet level. Collective action in lift irrigation does not show any visible concentration.

From the foregoing discussion, it is evident that locational factors do assume some importance in constraining the institutional choice of farmers. Thus, any explanation on farmers' choice should take into consideration the locational asymmetries as well.

Land and Crop Characteristics

Land and crop characteristics are also acknowledged as factors that are crucial in constraining the choice of farmers. The main variables that are relevant in this regard are land type, size of

Table 5.4: Distribution of Institutions according to Elevation

Institution	*Elevation from the outlet level (In metres)*								
	-4 to -3	-3 to -2	-2 to -1	-1 to -0.5	-0.5 to 0	0 to 0.5	0.5 to 1	1 and above	Total
Public Provision	17 (29.31)	16 (27.59)	8 (13.79)	5 (8.62)	4 (6.90)	5 (8.62)	2 (3.45)	1 (1.72)	58 (100.00)
Government Initiated Collective Action	—	1 (16.67)	1 (16.67)	3 (50.00)	1 (16.67)	—	—	—	6 (100.00)
Farmer Initiated Collective Action	4 (4.71)	6 (7.06)	20 (23.53)	22 (25.88)	11 (12.94)	10 (11.76)	12 (14.11)	—	85 (100.00)
Collective Action in Lift Irrigation	—	2 (7.69)	3 (11.54)	4 (15.38)	6 (23.08)	5 (19.23)	4 (15.38)	2 (7.69)	26 (100.00)
Water Market	—	—	—	—	3 (6.00)	23 (46.00)	18 (36.00)	6 (12.00)	50 (100.00)
Private Investment	—	6 (3.49)	14 (8.14)	8 (4.65)	14 (8.14)	36 (20.93)	50 (29.07)	44 (25.58)	172 (100.00)
Total	21 (5.29)	31 (7.80)	46 (11.59)	42 (10.58)	39 (9.82)	29 (7.30)	86 (21.66)	53 (13.35)	397 (100.00)

Note : Figures in brackets are percentages to total.
Source : Survey Data.

plot, type of cropping, and main and subsidiary crops cultivated.

(a) Land Type : Plots in the study area could be broadly classified into wet and dry. Canal and canal-based collective actions are found higher in the wetland, while dry land shows higher incidence of private investment, water markets and collective action in lift irrigation. Thus, 74 per cent of canal and canal-related institutions are seen in the wetlands; whereas about 86 per cent of all other institutions are in the dry lands. (Table 5.5). The following inferences can be arrived at from these results. The higher incidence of canal and collective action institutions, which are directly based on canal networks, in the wetlands indicates that the surface irrigation network could meet the requirements of wetland farming (basically paddy). The dominance of non-paddy mixed cropping in dry lands, which in turn demand diverse and timely irrigation, could be the reason for the increased prevalence of water markets, private investments and collective action in lift irrigation in the dry lands.

Table 5.5 : Distribution of Institutions across Land Type

Institution	*Land Type*		
	Wet	Dry	Total
Public Provision	43 (74.14)	15 (25.86)	58 (100.00)
Government Initiated Collective Action	3 (50.00)	3 (50.00)	6 (100.00)
Farmer Initiated Collective Action	64 (75.29)	21 (24.71)	85 (100.00)
Collective Action in Lift Irrigation	7 (26.92)	19 (73.08)	26 (100.00)
Water Market	3 (6.00)	47 (94.00)	50 (100.00)
Private Investment	25 (14.53)	147 (85.47)	172 (100.00)
Total	145 (36.52)	252 (63.48)	397 (100.00)

Note : Figures in brackets are percentages to total.
Source : Survey Data.

(b) Size of Holdings : Size of plots is generally perceived to be important in defining farmers' irrigation practices. However, the data does not show any discernible pattern to substantiate a significant correlation. All the institutions are scattered more or less in a similar manner without any specific concentration in holdings of any particular size. This may be due to the low dispersion of size holdings. A large proportion (72 per cent) of holding fall within the size class, 51-200 cents, as seen in (Table 5.6).

Table 5.6 : Distribution of Institutions across Size-holdings

Institution	Size of the Plot (In cents)						
	1-50	50-100	100-200	200-300	300-500	500 and above	Total
Public Provision	9 (15.52)	34 (58.62)	13 (22.41)	1 (1.72)	1 (1.72)	—	58 (100.00)
Government Initiated Collective Action	—	3 (50.00)	3 (50.00)	—	—	—	6 (100.00)
Farmer Initiated Collective Action	13 (15.29)	46 (54.12)	18 (21.18)	6 (7.06)	2 (2.35)	—	85 (100.00)
Collective Action in Lift Irrigation	7 (26.92)	5 (19.23)	12 (46.15)	1 (3.85)	1 (3.85)	—	26 (100.00)
Water Market	17 (34.00)	18 (36.00)	12 (24.00)	1 (2.00)	1 (2.00)	1 (2.00)	50 (100.00)
Private Investment	40 (23.26)	71 (41.28)	48 (27.91)	9 (5.23)	2 (1.16)	2 (1.16)	172 (100.00)
Total	84 (21.16)	177 (44.58)	107 (26.95)	19 (4.79)	7 (1.76)	3 (0.76)	397 (100.00)

Note : Figures in brackets are percentages to total.
Source : Survey Data.

(c) Cropping Pattern : Cropping pattern followed in the plots is also considered as a crucial variable in influencing the irrigation decisions. Crop type (mono *versus* mixed), as a variable constraining the institutional choice, shows results that are similar as in the case of land type. Mono-crop farming is characterised by higher incidence of canal-based institutions (Table 5.7), whereas with mixed cropping, there is a prominence of alternative institutions. This strengthens the argument that, canal irrigation and improvisations on it (collective actions) are preferred more for mono-crop (predominantly paddy) while mixed cropping warrants alternative institutions due to diverse water requirements.

Table 5.7 : Distribution of Institutions across Crop-type

Institution	*Crop Type*		
	Mono	Mixed	Total
Public Provision	41 (70.69)	17 (29.31)	58 (100.00)
Government Initiated Collective Action	3 (50.00)	3 (50.00)	6 (100.00)
Farmer Initiated Collective Action	64 (75.29)	21 (24.71)	85 (100.00)
Collective Action in Lift Irrigation	7 (26.92)	19 (73.08)	26 (100.00)
Water Market	10 (20.00)	40 (80.00)	50 (100.00)
Private Investment	23 (13.37)	149 (86.63)	172 (100.00)
Total	148 (37.28)	249 (62.72)	397 (100.00)

Note : Figures in brackets are percentages to total.
Source : Survey Data.

The profitability of the crop is also important in explaining the relationship between crop pattern and institutions existing. It is more likely that farmers engaged in the cultivation of non-paddy crops would be able to incur a higher cost and hence opt for alternative institutions, given the substantial difference in profitability between paddy and non-paddy crops. This is confirmed by the results of survey data, which shows an increased incidence of alternative institutions in non-paddy plots (Table 5.8).

On the whole, it seems that the possibility of farmers going for alternative institutions are higher when the crop is other than paddy. Out of 124 plots growing paddy, only 26 have alternative institutions other than canal-based systems. For crops such as coconut, plantain and vegetables alternative institutions are mostly resorted to. Out of 238 plots under coconut cultivation, only 47 plots are dependant on canal-based systems. This preference for alternative institutions in non-paddy plots could be attributed to the higher profitability of these crops.

Table 5.8 : Distribution of Institutions according to Cropping Pattern

Institution	*Main crop*						
	Paddy	Coco-nut	Nut-meg	Plan-tain	Are-canut	Vege-tables and Others	Total
Public Provision	39 (67.24)	19 (32.76)	—	—	—	—	58 (100.00)
Government Initiated Collective Action	3 (50.00)	3 (50.00)	—	—	—	—	6 (100.00)
Farmer Initiated Collective Action	56 (65.88)	25 (29.41)	—	3 (3.53)	—	1 (1.18)	85 (100.00)
Collective Action in Lift Irrigation	6 (23.08)	19 (73.08)	1 (3.85)	—	—	—	26 (100.00)
Water Market	—	25 (50.00)	—	16 (32.00)	—	9 (18.00)	50 (100.00)
Private Investment	20 (11.63)	147 (85.47)	—	1 (0.58)	1 (0.58)	3 (1.74)	172 (100.00)
Total	124 (31.23)	238 (59.95)	1 (0.25)	20 (5.04)	1 (0.25)	13 (3.27)	397 (100.00)

Note : Figures in brackets are percentages to total.
Source : Survey Data.

Institutions based on canal systems are prominent in paddy farming, followed by plots with coconut. Majority of the plots under public provision (67 per cent), 50 per cent under government initiated collective action and 66 per cent under farmer initiated collective action cater to paddy farming. Collective action in lift irrigation and private investment are prominently in plots with coconut. Of the total 26 plots under collective action in lift irrigation, 19 (73.08 per cent) had coconut as the principal crop. In the case of private investment, plots under coconut accounted for 85 per cent of the total plots. Water market is completely absent in paddy. This institution is prominent in plots with coconut, plantain and vegetables as main crops. The distribution of plots under water market across these crops is 50 per cent, 32 per cent and 18 per cent, respectively.

Apart from the main crop, the subsidiary crop could also be assumed to be important, as a large proportion of the plots studied fall under mixed-crop farming. In the case of coconut plots, in particular, a farmer's irrigation decisions are more likely to be influenced by the inter-crops such as plantain, nutmeg and vegetables, which are highly water sensitive. Thus, the analysis of the importance of crop will be complete only if the influence of subsidiary crops is also included. Accordingly, main subsidiary crop is examined as a constraint.

In the case of subsidiary crops also, there exists a clear correlation between crops grown and irrigation institutions chosen. The data on subsidiary crops in non-paddy plots indicate higher incidence of plantain, nutmeg and arecanut as the prime subsidiary crops (Table 5.9).

Table 5.9 : Distribution of Institutions according to Main Subsidiary Crop in Non-paddy Plots

Institution	*Main Subsidiary Crop*					
	Coco-nut	Nut-meg	Plan-tain	Are-canut	Vege-tables and Others	Total
Public Provision	2 (20.00)	—	8 (80.00)	—	—	10 (100.00)
Government Initiated Collective Action	1 (100.00)	—	—	—	—	1 (100.00)
Farmer Initiated Collective Action	1 (5.00)	—	19 (95.00)	—	—	20 (100.00)
Collective Action in Lift Irrigation	—	8 (44.44)	8 (44.44)	1 (5.56)	1 (5.56)	18 (100.00)
Water Market	1 (2.56)	7 (17.95)	27 (69.23)	3 (7.69)	1 (2.56)	39 (100.00)
Private Investment	3 (2.29)	60 (45.80)	29 (22.14)	24 (18.32)	15 (11.45)	131 (100.00)
Total	8 (3.65)	75 (32.25)	91 (41.55)	28 (12.79)	17 (7.76)	219 (100.00)

Note : Figures in brackets are percentages to total.
Source : Survey Data.

The plots under these subsidiary crops are mostly irrigated by institutions such as private investments, water markets or collective action in lift irrigation. All these, along with the fact that coconut accommodates mixed cropping of plantain, nutmeg, and arecanut lead to the inference that there is a higher chance of a farmer opting for alternative institutions (other than canal-based systems), given a mixed crop pattern where coconut is the main crop.

Personal and Household Characteristics of the Farmers

Personal characteristics of the farmers are also generally perceived as decisive factors in influencing the choice of institutions. The occupation, age and education of the household-head and the size of the family are the variables analysed below.

(a) Occupation of the Household Head : The following inferences can be drawn from the analysis of institutions across the main occupational categories of the respondents (Table 5.10). Farming is found to be the major occupation in the study

Table 5.10 : Distribution of Institutions across Occupation of the Household Head

Institution	*Main Occupation*				
	Culti-vator	Non-farm Work	Self-employed	Service and Others	Total
Public Provision	27 (46.55)	12 (20.69)	13 (22.41)	6 (10.34)	58 (100.00)
Government Initiated Collective Action	6 (100.00)	—	—	—	6 (100.00)
Farmer Initiated Collective Action	55 (64.71)	11 (12.94)	9 (10.59)	10 (11.76)	85 (100.00)
Collective Action in Lift Irrigation	13 (50.00)	2 (7.69)	6 (23.08)	5 (19.23)	26 (100.00)
Water Market	41 (82.00)	4 (8.00)	3 (6.00)	2 (4.00)	50 (100.00)
Private Investment	64 (37.21)	43 (25.05)	14 (8.14)	51 (29.65)	172 (100.00)
Total	206 (51.89)	72 (18.14)	45 (11.34)	74 (18.64)	397 (100.00)

Note : Figures in brackets are percentages to total.
Source : Survey Data.

area accounting for 52 per cent of the total farmers. This also points to the fact that about half of the farmers are part-time cultivators. About 60 per cent of the plots with canal-based systems have cultivation as the main occupation. The choice of private investment is higher among non-cultivators, especially those engaged in non-farm work and services. Around 60 per cent of those engaged in non-farm work and 69 per cent of service sector categories are found relying on private investment. This pattern could be due to the fact that private investment ensures more certainty and flexibility in irrigation given the part-time farmer status of the irrigators.

(b) Age of the Farmer : Age of the farmer was also analysed as it could also possibly determine the choice of farmers. However, no specific pattern was evident from the data. Farmers were mostly found in the age category of above 40 (Table 5.11). This age-wise homogeneity may be a reason for the absence of any definite pattern.

Table 5.11 : Distribution of Institutions across Age of the Household Head

Institution	*Age of the Farmer*				
	Upto 30	30-40	40-60	60 and above	Total
Public Provision	1 (1.72)	12 (20.69)	23 (39.66)	22 (37.93)	58 (100.00)
Government Initiated Collective Action	—	—	5 (83.33)	1 (16.67)	6 (100.00)
Farmer Initiated Collective Action	—	23 (27.06)	35 (41.18)	27 (31.76)	85 (100.00)
Collective Action in Lift Irrigation	—	8 (30.77)	14 (53.85)	4 (15.38)	26 (100.00)
Water Market	—	13 (26.00)	17 (34.00)	20 (40.00)	50 (100.00)
Private Investment	1 (0.58)	38 (22.01)	73 (42.44)	60 (34.88)	172 (100.00)
Total	2 (0.50)	94 (23.68)	167 (44.07)	134 (33.75)	397 (100.00)

Note : Figures in brackets are percentages to total.
Source : Survey Data.

(c) Size of the Family and Education of the Farmer : As in the case of the age of the farmers, size of the family and education of the farmer also do not show any clear pattern which could lead to a definite conclusion (Tables 5.12 and 5.13).

Table 5.12 : Distribution of Institutions across Family Size

Institution	*Size of the Family*			
	Less than 4	4-6	More than 6	Total
Public Provision	8 (13.79)	40 (68.97)	10 (17.24)	58 (100.00)
Government Initiated Collective Action	—	6 (100.00)	—	6 (100.00)
Farmer Initiated Collective Action	10 (11.76)	69 (81.18)	6 (7.06)	85 (100.00)
Collective Action in Lift Irrigation	2 (7.69)	15 (57.69)	9 (34.62)	26 (100.00)
Water Market	19 (38.00)	28 (56.00)	3 (6.00)	50 (100.00)
Private Investment	6 (3.49)	151 (87.79)	15 (8.72)	172 (100.00)
Total	45 (11.34)	309 (77.83)	43 (10.83)	397 (100.00)

Note : Figures in brackets are percentages to total.
Source : Survey Data.

In the case of both these variables, the patterns of concentration of households in specific ranges are more or less reflective of the general socio-demographic trends in the State. The analysis of the personal characteristics, thus, has shown that except for the main occupation of the farmer, none of the personal and household characteristics are found constraining the institutional choice.

On the whole, the analysis of the distribution of plots under various institutions across possible determinants shows that a number of variables could have influenced the decisions of farmers' choice of institutions. Of these, land, crop and locational factors seem to be vital. However, a definite conclusion in this regard could be made only after examining the statistical significance of these variables on the choice of institutions. This is dealt within the next section.

Table 5.13 : Distribution of Institutions across Education of the Household Head

Institution	*Completed Years of Education*				
	Less than 10	10-12	12-15	15 and above	Total
Public Provision	17	22	18	1	58
	(29.31)	(37.93)	(31.03)	(1.72)	(100.00)
Government Initiated Collective Action	1	2	1	2	6
	(16.67)	(33.33)	(16.67)	(33.33)	(100.00)
Farmer Initiated Collective Action	24	51	8	2	85
	(28.24)	(60.00)	(9.41)	(2.35)	(100.00)
Collective Action in Lift Irrigation	3	10	13	—	26
	(11.54)	(38.46)	(50.00)		(100.00)
Water Market	15	23	12	—	50
	(30.00)	(46.00)	(24.00)		(100.00)
Private Investment	32	99	37	4	172
	(18.60)	(57.56)	(21.51)	(2.33)	(100.00)
Total	92	207	89	9	397
	(23.17)	(52.34)	(2.01)	(2.27)	(100.00)

Note : Figures in brackets are percentages to total.
Source : Survey Data.

Multinomial Logit Model of Institutional Choice

The statistical significance of the aforesaid variables is verified using a multinomial logit model[2]. The method is appropriate for studying the relationships between a number of covariates and a dependent variable that has more than two possible outcomes. Besides, the model is used for predicting the probabilities of incidence of various outcomes.

The role of various factors in determining the institutional choice of farmers is examined in the multinomial logit framework. In the model, the relative effects of three sets of variables (locational, crop, land and household characteristics) in the institutional choice of the farmers are analysed. Further, it also predicts the probability of choice of various institutions in the provision of irrigation.

In the multinomial logit, a set of coefficients say, $\beta^{(1)}$, $\beta^{(2)}$, and $\beta^{(3)}$ corresponding to each outcome category such that :

$$P(y=1)=\frac{e^{x\beta(1)}}{e^{x\beta(1)}+e^{x\beta(2)}+e^{x\beta(3)}+e^{x\beta(4)}+e^{x\beta(5)}+e^{x\beta(6)}}$$

$$P(y=2)=\frac{e^{x\beta(2)}}{e^{x\beta(1)}+e^{x\beta(2)}+e^{x\beta(3)}+e^{x\beta(4)}+e^{x\beta(5)}+e^{x\beta(6)}}$$

$$P(y=3)=\frac{e^{x\beta(3)}}{e^{x\beta(1)}+e^{x\beta(2)}+e^{x\beta(3)}+e^{x\beta(4)}+e^{x\beta(5)}+e^{x\beta(6)}}$$

The model however, is unidentified in the sense that there is more than one solution to the $\beta^{(1)}$, $\beta^{(2)}$, and $\beta^{(3)}$ that leads to the same probabilities for y=1, y=2, and y=3. To identify the model one of the $\beta^{(1)}$, $\beta^{(2)}$, or $\beta^{(3)}$ is set to zero (base category selection). If $\beta^{(1)}$ is set to 0, then the remaining coefficients $\beta^{(2)}$, and $\beta^{(3)}$ would measure the change relative to y=1 group. If $\beta^{(2)}$ is set to 0 the remaining coefficients $\beta^{(1)}$, and $\beta^{(3)}$ would measure the change relative to the y =2 group. The coefficients would differ because they have different interpretations, but the predicted probabilities for y=1, 2, 3 would be the same. Thus, either parameterization would be a solution to the underlying model.

Setting $\beta^{(1)} = 0$, the equations for estimating the probabilities become :

$$P(y=1)=\frac{1}{1+e^{x\beta(2)}+e^{x\beta(3)}+e^{x\beta(4)}+e^{x\beta(5)}+e^{x\beta(6)}}$$

$$P(y=2)=\frac{e^{x\beta(2)}}{1+e^{x\beta(2)}+e^{x\beta(3)}+e^{x\beta(4)}+e^{x\beta(5)}+e^{x\beta(6)}}$$

$$P(y=3)=\frac{e^{x\beta(3)}}{1+e^{x\beta(2)}+e^{x\beta(3)}+e^{x\beta(4)}+e^{x\beta(5)}+e^{x\beta(6)}}$$

The relative probability of y = 2 to the base category is : *Pr (y=2)/Pr(y=1)* = $e^{x\beta(2)}$

Model Specification and Rationale of Variable Selection

The model postulates that the probability of a farmer going for a particular institution depends upon his attributes classified as follows. Dependent variable (INST) defines the institutional choice of the farmers. In CRDS as explained in Chapter IV, there are six irrigation institution outcomes such as public provision, government initiated collective action, farmer initiated collective action, collective action in lift irrigation, water markets and private investment. Considering some common characteristics of these institutions for the purpose of analysis in the model, these are broadly categorized into three sets—*public provision; collective action in canal;* and *other institutions*. Here, *collective action in canal* includes both government initiated and farmer initiated collective actions and *other institutions* include : collective action in lift irrigation, water markets and private investment. Accordingly, the dependent variable INST takes three values : INST=1 for public provision; INST=2 collective action institutions in canal; INST=3 other institutions. INST=1 (public provision) is taken as the base category in estimating the multinomial logit, as all institutions exist in the canal command and canal water forms the major source of most of these alternative institutions.

There are three sets of independent variables. The first set consists of locational variables and it captures the influence of the position of a plot in deciding the institutional choice. The inclusion of these variables was considering the advantages/disadvantages of irrigators depending on the position of the plots in the canal, a widely acknowledged finding in the literature. Empirical studies on cooperation are often suggestive of the variation in possibilities of collective action with respect to the changes in the reach of the farm (head/middle/tail ends). There are four variables in the set namely : distance from the main canal (DISTM), distance from the branch canal (DISTB), distance from the outlet (DISTO) and the elevation of the plot (ELEV). DISTM is further divided into two categories : upto 10 km; and more than 10 km. Likewise, DISTB, DISTO and ELEV are also categorized into sub-groups. These covariates are all categorized considering the frequency distribution of the plots across locational coordinates.

The second set of variables, land and crop characteristics are intended to capture the role of farm characteristics in deciding the institutional selection. The variables in this set are : size of holding (LSIZE); type of land (LTYPE); and main crop (CMAIN). The inclusion of LSIZE in the model is due to well-acknowledged positive correlation between farm size and ownership of agricultural assets. The probability of going for alternative institutions, which require larger investments, is expected to be higher among large holders. LSIZE is divided into two categories : upto 1 acre and above 1 acre to capture the effect across small and medium/large farms. LTYPE is a dummy which denotes whether the land is wet or not (LTYPE=1, if wet land; LTYPE = 0, if dry land). As the extent of water availability varies between wet and dry land, one can expect a higher possibility of dry land plots going for alternative institutions, with higher certainty in terms of water availability. CMAIN denotes the main crop of the plot and is again a dummy that takes the value of 1, if paddy is the main crop; and 0, otherwise. A greater incidence of canal and canal-related institutions are expected when paddy is the main crop. It is reasonable to assume that non-paddy plots with mixed crop pattern are more likely to opt for those institutions, which are suitable for diverse water requirements and ensure increased water control. Further, in the context of the study, the fact that the whole canal system is designed to cater to paddy farming adds to the importance of analysing the significance of the variable.

The third set includes household characteristics with variables such as : main occupation of the owner of the plot (OCCU), age (AGE), education (EDU), and size of the family (FSIZE). OCCU is taken as a dichotomous variable with value 1, if cultivator; and 0, otherwise. As part-time cultivators do not have much time for farm activities, there is a higher chance for choosing institutions that require less time in transacting water and provide more flexibility in irrigation timings. Thus, it is reasonable to expect a higher incidence of private investment among non-cultivators. Age (AGE) and education

(EDU) are also found to be influential as experience and learning could effect institutional selection. Both these covariates are classified into two categories—up to 45 and above 45 in the case of AGE; and upto 10 years of education and more than 10 years in the case of EDU. Family-size (FSIZE) could also affect the institutional choice as increased availability of family labour may lead farmers to choose institutions that demand more time and effort to transact water. Higher incidence of collective action institutions is observed, in empirical studies, with increased size of households. FSIZE is divided into two sub-categories such as : up to 4 members and more than 4 members.

Estimation Results

(a) Probabilities of Institutional Choice : The estimates of probability of institutional choice have shown a higher probability for other institutions (0.62) followed by collective action institutions (0.22) and canal irrigation (0.15). The results reveal the clear domination of alternative irrigation institutions in the canal command.

(b) Relative Contribution of Various Factors : The results of the estimated multinomial logit model indicate the factors influencing the choice of irrigation institutions in the context of CRDS and the coefficient estimates are given in Table 5.14.

The analysis of the effect of locational factors on institutional choice reveals that these factors are significant in the choice of both collective action in canal and other institutions. In the case of *collective action institutions in canal,* the effect of DISTM, DISTB and DISTO are found significant and negative indicating the lower chances of opting the institution beyond 10 km from the main canal, 2 km from the branch canal and 0.5 km from the outlet, respectively. The results are mostly in the expected direction, confirming higher probability of collective action institutions (relative to public provision) in plots, that are not very far from the main and branch canals, and located near to the outlets.

Table 5.14 : Estimates of Multinomial Logit Model on Determinants of Institutional Choice

Variable	*Collective Action in Canal/Public Provision*		*Other Institutions/ Public Provision*	
	Odd Ratios	SE	Odd Ratios	SE
DISTM				
Upto 10 km ®	1.00		1.00	
>10 km	0.39	0.23**	1.15	0.51*
DISTB				
Upto 2.00 km ®	1.00		1.00	
> 2 km	0.55	0.15**	1.24	0.61***
DISTO				
Upto 0.50 km ®	1.00		1.00	
> 0.5 km	0.89	0.44*	0.81	0.45**
ELEV				
Upto outlet level ®	1.00		1.00	
Above outlet level	1.49	0.63	5.31	2.46***
LSIZE				
Upto 1 acre ®	1.00		1.00	
> 1 acre	0.44	0.57	8.63	15.64
LTYPE				
Dry ®	1.00		1.00	
Wet	1.37	1.16	0.10	0.09***
CMAIN				
Paddy ®	1.00		1.00	
Non-paddy	0.19	0.14**	1.21	0.58**
OCCU				
Non-cultivator ®	1.00		1.00	
Cultivator	0.58	0.26	0.22	0.11***
FSIZE				
Upto 4 members ®	1.00		1.00	
> 4 members	1.09	0.42	1.04	0.44
EDU				
Upto 10 years ®	1.00		1.00	
>10 years	2.13	0.83**	2.40	1.04**
AGE				
Upto 45 years ®	1.00		1.00	
> 45 years	1.26	0.54	2.87	1.41
N		397		
Log likelihood		−233.93		
Model chi-square		256.74		

(*INST*-1=public provision, 2=collective action institutions in canal, 3=other institutions; base category, 1=public provision).

Note : ®—Reference category; *** *p<0.01*, ** *p<0.05*, * *p<0.10*.

Source : Survey Data.

In the case of *other institutions*, prominence of significant effects is observed for all locational variables. DISTM, DISTB and ELEV are found to be positively related indicating higher chances of *other institutions* (outside canal) in distant locations from the canal network, where water is scarce and uncertain. The direction of effect of DISTO was found negative, which could be attributed to the lower likelihood of farmers incurring considerable investment, when the possibility of getting seepage water[3] is unsure.

As regards the land and crop characteristics, CMAIN is significant in both collective action and other institutions. In the case of collective action (in canal), CMAIN is negatively related indicating lower chances of non-paddy growers opting for canal-based institutions, than public provision. However, in the case of other institutions, the effect of CMAIN is positive. This is also along predicted lines, as non-paddy growers are more likely to go for institutions that suit the diverse water requirements of these plots, which are largely with mixed cropping. This is indicative of the fact that farmers are ready to go for alternative irrigation institutions outside canal, if the main crop cultivated is non-paddy. The relatively higher profitability from non-paddy crops could be the main driving force behind this. LTYPE is significantly and negatively related in the case of wet land revealing lower chances of other institutions in wet land plots. This could be attributed to the fact that being low-lying areas, water shortage is comparatively not severe in these plots. Further, as these plots are predominantly under paddy farming there is not much mismatch between the water requirement and supply as the canal system is primarily designed for irrigating paddy.

Among the household characteristics, EDU is found significant and positive in both the cases of collective action in canal and other institutions, which shows higher preferences for alternative institutions among educated farmers. OCCU (cultivation as main occupation) is negatively and strongly significant in the choice of other institutions, which explains the strong preference among full time cultivators not to go for other institutions. Conversely, it suggests the higher chances among part-time cultivators to go for other institutions with increased flexibility and water control. This relation is expected, as canal related institutions require considerable time and effort in organising and negotiating.

From the foregoing discussion of the determinants of institutional choice, it is evident that the role of different variables varies across institutional categories. Though the importance of locational variables is striking, it is significant that no definite pattern is emerging due to the variance in the significance of different factors across institutional categories. As is clearly evident from the data the plots with paddy are largely found depending on canal-based irrigation, while non-paddy plots are mostly with alternative institutions. However, even within the categories of paddy plots and non-paddy plots, the choice of institution does vary considerably across plots. This is especially so, across various canal-based institutions in the case of paddy plots and within alternative institutions in the case of non-paddy plots. Keeping this aspect in view, it is reasonable to attempt a disaggregate analysis of determinants of institutional choice for plots with distinct crop pattern (paddy *versus* non-paddy) which could provide a closer understanding of the roles of the other factors (locational and personal characteristics) in the institutional choice. This would also help in controlling the effect of crop characteristics in the determination of institutional choice.

Determinants of Institutional Choice across Crop Categories

In the CRDS command area, as has been discussed in Chapter IV, the crops cultivated are broadly of two types; paddy and non-paddy. Non-paddy plots are mostly under mixed cropping with coconut as the main crop along with, nutmeg, plantain or arecanut as main subsidiary crop[4]. To capture the determinants, multinomial logit analysis is attempted for those with paddy farming (124). However, for non-paddy plots, a logistic regression model was estimated as the number of plots under public provision with non-paddy crop was found too small for estimating the multinomial logit model. As the division of the sample was based on crop grown, crop related variable, LTYPE—wet or dry is also not taken in the model.

Estimation Results for Paddy Plots

(a) Probabilities of Institutional Choice : The estimates of probability of institutional choice have shown equal probability for canal irrigation and collective action institution (0.40). Other institutions showed a comparatively lower probability (0.20). The results are indicative of the prominence of public provision and other canal-related institutions in paddy plots.

(b) Relative Contribution of Various Factors : The results of the estimated multinomial logit model indicate the factors influencing the choice of irrigation institutions in the paddy plot (Table 5.15). The results reveal that locational factors are important in the choice of institutions, given paddy as the main crop. DISTM, DISTB and ELEV are found significant in influencing farmers' choice for *collective action institutions in canal*. The effects of DISTM, DISTB and ELEV are found negative, beyond certain distance and elevation (10 km, 2 km and 0.5 metre respectively) indicating the lower chances of collective action in canal if the plots are located away from the canal network. This is in conformity with the oft-cited argument that collective action institutions are difficult to sustain in the tail ends and high-land plots, where acute water scarcity exists. In other words, farmers prefer not to put collective efforts if the scope for improving water is limited.

In the case of *other institutions* also the locational factors are found significant and influential. DISTM and DISTB are negatively related while ELEV shows a strong positive relation. The results suggest that the existence of other institutions is prominent in plots, which are close to the main canal and branch canal. However, this does not seem to suggest that other institutions in paddy plots are more likely in locationally advantageous positions, as ELEV show a strong positive relation. This would mean that other institutions in paddy plots are more likely in plots located at a higher elevation from the outlet but are not far away from the canal network.

Table 5.15 : Estimates of Multinomial Logit Model on Determinants of Institutional Choice in Paddy Plots

Variable	*Collective Action in Canal/Public Provision*		*Other Institutions/ Public Provision*	
	Odd Ratios	*SE*	*Odd Ratios*	*SE*
DISTM				
Up to 10 km ®	1.00		1.00	
>10 km	0.28	0.19*	0.04	0.06***
DISTB				
Up to 2.00 km ®	1.00		1.00	
>2 km	0.62	0.31**	0.09	0.09***
DISTO				
Up to 0.50 km ®	1.00		1.00	
>0.5 km	1.30	0.75	0.14	0.19
ELEV				
Up to outlet level ®	1.00		1.00	
Above outlet level	0.88	0.51**	24.21	19.39***
LSIZE				
Up to 1 acre ®	1.00		1.00	
>1 acre	0.85	1.25	0.80	0.55
OCCU				
Non-cultivator ®	1.00		1.00	
Cultivator	0.55	0.31	0.16	0.12
FSIZE				
Up to 4 members ®	1.00		1.00	
>4 members	1.42	0.65	2.45	1.77
EDU				
Up to 10 years ®	1.00		1.00	
>10 years	2.22	1.01*	4.23	3.17
AGE Up to 45 years ®	1.00		1.00	
>45 years	2.06	1.09	1.55	4.40
N		124		
Log likelihood		–98.35		
Model chi-square		65.18		

(*INST*-1= public provision; 2=collective action institutions in canal; 3=other institutions; base category, 1=public provision)

Note : ®—Reference category; *** *p<0.01,* ** *p<0.05,* * *p<0.10.*

Source : Survey Data.

Among the other variables only EDU is significant indicating stronger preference among educated farmers to go for collective action. This could be attributed to the fact that, a major chunk of the collective action in CRDS falls under CADA or group farming (both Government-based), which requires constant liaisoning and interface with government officials.

Estimation Results for Non-paddy Plots

The number of non-paddy plots under public provision was found too small (9 out of 273 cases) to estimate the relationships of other categories with respect to public provision using the multinomial logit model. Hence, a logistic regression was run out to examine the effects of various factors in determining the institutional choice, after removing the cases of public provision. Logistic regression is considered more appropriate in this case, as after the elimination of public provision, the dependent variable became dichotomous (collective action *versus* other institutions). The Table 5.16 shows the results of the logistic regression for non-paddy plots.

The results for non-paddy plots also confirm the importance of locational variables in choosing the institution, as has been seen in the case of paddy plots. All the locational variables are found to have positive and significant influence on the choice of other institutions, which shows the tendency of farmers to depend on private initiatives (compared to collective action) when the plots are disadvantageously placed. Apart from locational variables, OCCU is significant and negative indicating lower chances of cultivators to go for private initiatives compared to institutions of collective action.

The foregoing analysis, aggregate as well as those separately for paddy and non-paddy plots, leads to the conclusion that locational variables are highly significant in influencing the institutional choice of farmers. As discussed in Chapter III, farmers' choice of institutions from the set of multiple institutions depends on the expected pay-offs from various institutions. As benefits of institutional choice are more or less a function of crops grown, it could be assumed that benefit calculations would not arise as the profitability of farming remains almost the same across all the plots with the same crop pattern. Hence, for plots with a given crop, institutional choice could be assumed to be a function of cost minimization, given the constraints. Accordingly, the institutional choice in CRDS could be analysed as a cost minimisation strategy of the farmers, constrained mainly by locational factors. These cost considerations in institutional choice are discussed in the following chapter.

Table 5.16 : Estimates of Logistic Regression on Determinants of Institutional Choice in Non-paddy Plots

(*INST*-1=other institutions; 0=collective action institutions in canal)

Variable	*Odd Ratios*	*SE*
DISTM		
Up to 10 km ®	1.00	
>10 km	2.12	0.22 **
DISTB		
Up to 2.00 km ®	1.00	
>2 km	2.07	0.92***
DISTO		
Up to 0.50 km ®	1.00	
>0.5 km	1.79	0.45**
ELEV		
Up to outlet level ®	1.00	
Above outlet level	2.56	0.99**
LSIZE		
Up to 1 acre ®	1.00	
>1 acre	2.16	1.86
OCCU Non-cultivator ®	1.00	
Cultivator	0.74	0.14*
FSIZE		
Up to 4 members ®	1.00	
>4 members	1.09	0.44
EDU		
Up to 10 years ®	1.00	
>10 years	1.17	0.46
AGE		
Up to 45 years ®	1.00	
>45 years	1.70	0.67
N	264	
Log likelihood	–95.02	
Model chi-square	37.95	

Note : ®—Reference category; *** *p<0.01,* ** *p<0.05,* * *p<0.10.*
Source : Survey Data 0.10.

NOTES

1. The source of water in water market institution is also mostly canal. However, the plots representing water markets in the analysis are buyers' plots.
2. Multinomial logit models are useful in situations, where the dependent variable takes more than two categories. The model helps in analysing the outcome as to how the actual decisions are made and the roles of various factors that are likely to affect such decisions in a probabilistic sense. (see Domenich and McFadden, 1975 and Greene, 1993, for details).
3. In CRDS, seepage water from canal is the major source of water in all the different institutions except in few cases with lift irrigation and water market.
4. The cropping pattern in the command area has remained more or less unchanged since 1980s which is marked with a significant decline in paddy cultivation, sharp increase in mixed crop (coconut, banana, nutmeg, arecanut and vegetables).

6

Analysis of Cost in Institutional Choice

Associated with various alternative institutions, there exist differential cost-benefit outcomes and the rational behaviour of farmers demands the selection of institution, in a net-gain maximising manner. The choice of a particular institution by a farmer, thus, reveals that the farmer's perception of net-gain associated with the institution is higher than that of all other alternatives. In irrigation, two considerations influence the calculations of net-gains leading to the choice of institution; cost minimisation and benefit maximisation. Benefit is more or less a function of crops grown. It is reasonable to assume that, given the cropping pattern, the benefit calculations would not arise in the institutional choice of irrigation. For a defined crop pattern, returns from farming remain almost the same for all the plots, and also the level and costs of other inputs such as fertilisers, seeds are also given. Under this condition, the net-gain maximisation behaviour of the farmers could be reduced to a strategy of minimising the cost of acquiring irrigation. The fact that different institutions exist in plots with same crop pattern points to the differential costs of institutions across plots. With this view, the costs associated with various institutions in CRDS are estimated.

Costs across Institutions

Costs across the different institutions are calculated from the survey data (Table 6.1). The average cost is divided into two components; fixed/sunk cost, and variable/recurrent costs. Fixed costs are the one-time, direct costs incurred by an individual farmer to organise an institution, while the variable costs are the direct recurring cost associated with the institution.

Table 6.1 : Average Costs across Institutions (In Rupees per acre)

Institution	*Average Fixed Cost*	*Average Annual Variable Cost*
Public Provision	—	—
Government Initiated Collective Action	—	16
Farmer Initiated Collective Action	154	104
Collective Action in Lift Irrigation	242	140
Water Market	96	540
Private Investment	5484	220

Source : Survey Data.

It is evident from the Table 6.1 that the fixed and variable costs vary across institutions. Public provision has no additional costs other than irrigation cess. (Cess on irrigation is not included in the cost, as it is common to all irrigators in the command area, irrespective of the institution, unless specific exemptions are given in some plots.) Private investment involves the highest fixed cost, whereas water markets entail maximum variable costs. Collective action in lift irrigation and private investment imply the second highest fixed and variable costs respectively.

The apparent difference in these costs associated with various institutions lead to an obvious question as to why farmers go for those institutions that entail higher cost? This demands a broader conceptualisation of costs associated with institutions, recognising the importance of transaction costs in the economic decisions of individuals.

Transaction Costs and Institutions

Transaction costs are defined as 'the costs of running the economic system' and are interpreted as 'the comparative cost of planning, adapting and monitoring task completion under alternative governance structures' (Arrow, 1969). North (1990) describes transaction costs as 'the costs of measuring the valuable attributes of what is being exchanged and the costs of protecting rights and policing and enforcing agreements'.

Any economic activity could be organised through different institutions, and implicit in any institution, there exist problems of contracting and associated procedures between the actors. Hence, positive costs exist in devising, enforcing, maintaining and monitoring institutions, which is termed as transaction costs. There are *ex ante* and *ex post* components of transaction costs. The former includes the cost of drafting, negotiating and forming a contract or agreement; the latter is concerned with the costs of enforcement and monitoring a contract or agreement, and search and information costs, including those of remedial measures for deviations from pre-specified agreements and contractual provisions (explicit and implicit), cost of misalignments and maladaptation of transactions that drift out of agreed specifications and alignment and costs of dispute resolution (Rao, 2003). Both these sets of costs are interdependent.

In irrigation, transaction costs could be conceived as the indirect or hidden costs associated with getting the institutions function. These costs do exist irrespective of the category of institution through which irrigation is organised. In irrigation, transaction costs mostly constitute the costs of identifying profitable opportunities for acquiring water (information costs), costs of negotiating and bargaining or administratively deciding on the water transfer, the institutional costs of actually organising and contracting (formal and informal) and monitoring the transfers, and the cost of monitoring, enforcing and mitigating or eliminating possible third party effects and externalities. The costs of search and information, haggling costs and waiting costs are also important components of transaction costs in irrigation. Transaction cost could vary not only across institutions but also across individuals, depending on their personal attributes. Individual's valuation of time is a major factor influencing transactions cost in irrigation. Thus, an institution that minimises production cost need not be the one that minimises total costs due to the difference in transaction costs. The foregoing discussion highlights the relevance of considering transaction cost also in analysing institutional choice. Hence, the cost associated with various irrigation institutions can be divided into two : (1) production costs, and (2) transaction costs. In this broader framework, the cost components across various irrigation institutions are discussed subsequently.

Cost Components across Irrigation Institutions

Within the canal command, all the irrigators are to pay a cess, which is fixed by the State. Apart from this, every farmer incurs a positive cost in the appropriation of irrigation, which vary across institutions and plots. The cost implications of various institutions in CRDS, as captured by the survey, are explained below.

(a) Costs in Public Provision : Theoretically, canal water is supplied to the farmer at a subsidised rate (i.e., the cess). The Government sets up the common facilities and is responsible for the delivery of water up to the outlet. No costs are to be incurred by the farmer above the outlet, legally. However, in practice, an individual farmer may incur several visible and hidden costs in *getting the supply right* at the outlet level. These costs include, the individual expenses incurred on liaisoning with the irrigation authorities (on travel and meeting officials), gathering information regarding water availability and turns, influencing and bribing canal lascars and officials, negotiating with competing farmers, and enforcing and monitoring the schedules. Beyond the outlet, the costs include individual expenses on construction and maintenance of channels, negotiating with neighbouring farmers, conflict resolution and organisation, monitoring and enforcing the turns.

(b) Costs in Government Initiated Collective Action : The major cost of an individual farmer in government initiated collective action is the individuals' share on organising the collective and construction of common facilities. These are the individual shares of the total cost on collective efforts, which include organisation and maintenance cost of the institution. Much of the one-time cost of organising (cost associated with the setting up of the institution, formulation of the rules and regulations etc.) is borne by the State under the CAD Programme. However, individuals do incur several hidden and direct costs in terms of time, money and efforts on information gathering, consensus building, organising and so on. At times, individual farmers may also have to share the production costs supplementing the government funds for constructing common facilities. Once, the institution is in place, the farmer's cost includes the expenses on maintaining the institution through persistent efforts at organising and coordinating, liaisoning with CAD officials, monitoring and enforcing of the rules and regulations. The farmer may also incur *external costs*, as a member of the group. These are individual costs involved in complying with the decisions taken in the interest of the group.

(c) Costs in Farmer Initiated Collective Action : The major cost implication for an individual farmer in farmer-initiated collective action is the contribution towards the organisational cost of the group. These include costs of information gathering, coordination, negotiation, decision-making and external costs. The organisational cost also involves the expenses on consensus building and that on developing rules for the functioning of the group. The individual's share of costs on developing rules for the group assumes importance and is essentially related to the cost of consensus-building on the norms of sharing water and the costs; monitoring and reporting mechanisms, instituting punitive actions for violations, mechanisms for conflict resolution and development of a governance structure. Once the group begins functioning, the individual farmer has to contribute towards the production of common facilities and their maintenance. Apart from this, several costs need to be incurred by the members regarding monitoring, ensuring compliance, resolving conflicts, liaisoning with irrigation officials, information gathering and collection of individual contributions. This also involves individual costs on continued consensus building and organising. External costs also assume importance, as the farmer may have to abide by the common interest of the group leading to personal losses.

(d) Costs in Collective Action in Lift Irrigation : The major individual cost in the case of collective action in lift irrigation relates to the organising cost of the system. This includes individual's share of costs on information gathering, negotiation and consensus building, influencing the authorities, setting up common facilities (installation of pump-sets, construction of reservoirs, and field channels) and devising of rules and regulations for sharing water and mechanisms for conflict resolution. In cases of government owned lift irrigation projects, these organisational costs are borne by the state. However, in community lift irrigation schemes though a proportion is borne by some community organisation a larger share is the farmers' contribution. Once the institution is in operation, the farmer's costs include water charges, individual contribution towards maintenance of the system and salary of the operators/watchmen and monitoring, enforcing and reporting. As collective action in lift irrigation involves group coordination, *external costs* are also present.

(e) Cost Associated with Water Market : The farmer has to incur an initial cost associated with organising and negotiating the institution. These costs mainly relates to information gathering, negotiation and contract formulation. Farmers also have to incur the costs on building up the watercourses (construction of earthen channels, purchase of PVC pipes). At times, these costs are minimal due to the presence of already-built infrastructure (for instance CAD field channels). The running costs of the institution for a buyer are mainly water charges and the recurring expenses on monitoring and enforcing the contract.

(f) Costs in Private Investment : The major cost in private investment is that associated with the setting up of physical infrastructure, (well/pond and pump-sets). In cases where subsidy is involved, the state bears a proportion of these installation costs. In these cases, the individual would have to incur costs on information gathering, negotiation and liaisoning with authorities. The recurring costs are mainly the fuel charges (diesel or electricity) and the maintenance charges of the physical infrastructure. Table 6.2 provides a summary of the constituents of costs across various institutions existing in CRDS canal command.

Table 6.2 : Constituents of Cost across Institutions

Institution	*Type of Cost*	*Production Cost*	*Transaction Cost*
Public Provision	Fixed	Nil	Nil
	Variable	Maintenance of channels in farmer's own plots	Information gathering about the turns, negotiation with neighbouring farmers and canal officials, conflict resolution, bribing of canal lascars or irrigation officials, influencing (both political and bureaucratic) and monitoring the implementation of turns.
Government Initiated Collective Action	Fixed	Individual's contribution (if any) to the construction of common facilities and field channels in addition to the state's contribution. In most of the cases the individual's contribution is negligible with state contributing the entire expenses.	Individual's contribution in information gathering (about the benefits, possibilities and modalities of the CAD Programme), consensus-building among the fellow irrigators, negotiations with CAD officials, and organisation (the scheduling of water turns, designing the rules and norms of water allocation, cost recovery and conflict resolution)
	Variable	Individuals' share in maintenance of channels and other common facilities	Time, Money and Resources spent by individuals in maintaining the organisation (continued efforts on consensus-building, negotiation, conflict resolution, and sorting out externalities). Farmers' contribution in terms of efforts in running the system (such as regulating and monitoring the turns).Individual's expenses on monitoring, regulating and reporting on the functioning of the institution.

(contd.)

(TABLE 6.2 contd.)

Institution	*Type of Cost*	*Production Cost*	*Transaction Cost*
Farmer Initiated Collective Action	Fixed	Individual's share on construction of common facilities—setting up of pump-set, digging or construction of reservoirs/wells, construction of water courses etc.	Farmer's share on information gathering (regarding water turns, possibilities of improving water supply, benefits of cooperation etc.), consensus-building among fellow irrigators, liaisoning with irrigation officials, negotiations and organisation (defining the course of action with specific rules and norms)
	Variable	Farmers' recurring contribution towards maintenance of channels and other common facilities	Individual share in terms of money, time and efforts in information gathering, consensus-building and negotiation, conflict resolution, collective bribing and influencing—both officials and politicians.Farmers' cost on monitoring, regulating, dealing externalities and reporting.Individual share (if any) in running the system—for instance, regulating the turns etc.
Collective Action in Lift Irrigation	Fixed	Farmers' contribution towards the construction of facilities in addition to the expenses borne by the state/other agencies	Individual farmers' share of costs incurred on information gathering, consensus building, negotiation and organisation
	Variable	Water charges, Individual share in the maintenance of facilities	Individual expenses (both monetary and labour) on conflict resolution, monitoring, externalities, regulating and reporting
Water Markets	Fixed	Cost incurred in construction of facilities to transact water through the institution—setting up of water courses, purchase and installation of PVC pipes etc.	Information gathering regarding the possibilities of the institution. Expenses on negotiation and formulation of the contract.

(contd.)

(TABLE 6.2 contd.)

Institution	*Type of Cost*	*Production Cost*	*Transaction Cost*
	Variable	Water charges, expenses on maintenance of facilities.	Expenses on enforcing and monitoring the contract. Expenses on maintaining the relation with the seller of water.
Private Investment	Fixed	Expenses on construction of facilities (sometimes subsidised by the government—subsidies on construction of wells and purchase of pump sets)	Expenses on gathering information on possibilities, organising loans and subsidies, and efforts on installing the system—on digging of wells, purchase and setting up of pumps, availing electric connection etc.
	Variable	Fuel charges (subsidised by the government if electricity), maintenance of facilities—repair of well, motor etc.	Nil

Source : Survey Data.

Given positive transaction costs associated with various institutional alternatives of acquiring irrigation water, the rational behaviour of the farmer suggest that economising takes place not with reference to production costs but with total costs, which is the sum of production and transaction costs. Thus, it could be argued that institutions with higher production costs could also exist, if it ensures minimum total costs.

Despite their importance, transaction costs are difficult to quantify. This is, however, mitigated by the fact that these are always assessed in a comparative institutional way, in which one mode of contracting is compared with another (Williamson, 1985). Accordingly, it is the difference between rather than the absolute magnitude of transactions costs that matter[1].

In this context, rather than quantifying the transaction costs across various institutions, it is attempted to analyse the cost minimisation considerations (which *inter alia* captures the transaction cost calculations) of farmers in institutional choice. As discussed in the previous chapter, the locational variables do act as major determinants in the choice of institutions. Due to the diversity of plots in terms of locational variables, different farmers (plots) may have to incur different costs under the same institution.

Locational Variables and Transaction Cost Implications

The cost associated with different institutions differs across plots. As locational variables are found to be important in determining irrigation institutions, any strategy of cost minimisation needs to be captured in terms of locational specificity of the plots. Though theoretically all the institutions are available for a farmer, there could only be one institution—the cost minimising solution for realising a particular supply of irrigation, within the constraints set by the location of the plots. Thus, the selection of a particular institution by a farmer in a plot would always mean that all the other alternatives are considered costlier in realising the same amount of water. An analysis of the locational specificities across various irrigation institutions could provide important insights into the transaction cost considerations and cost minimising strategies of individual irrigators. For this, it is imperative to delineate the specific combination of locational variables that shape an institutional choice. Accordingly, the mean values of all the locational variables are estimated and compared across institutions, to obtain the set of coordinates of locational factors that influences cost calculations of individuals leading to the selection of a particular institution. To avoid the issue of difference in benefit, the plots are divided into two sub-sets; non-paddy and paddy plots.

Locational Concentration of Institutions in Non-Paddy Plots

The mean value for locational variables for non-paddy plots (273) has been estimated and is given in Table 6.3. In the case of public provision, distances from the main and branch canals are found less than those of other institutions. The elevation of the plot is also found much below the outlet and has the lowest value among all institutions. Water availability from canal directly depends on the proximity of the plot to the main and branch canals, and hence those plots, which are positioned better off in terms of these two variables, are more likely to opt for public provision. Also, the availability of seepage water is also high, as the plots are located well below the outlet level. Thus, it is reasonable to argue that the choice of public provision is more likely in those plots, which are fairly near to the water flows in the canal and with scope for large amount of seepage.

Table 6.3 : Mean Values of Locational Variables across Institutions for Non-paddy Plots

Institution	*DISTM**	*DISTB**	*DISTO**	*ELEV***
Public Provision	10.00	1.65	0.38	–0.67
Govt. Initiated Collective Action	11.88	3.05	0.30	–0.43
Farmer Initiated Collective Action	12.45	2.06	0.24	0.10
Collective Action in Lift Irrigation	11.90	1.93	0.75	0.48
Water Market	11.72	2.07	0.45	0.81
Private Investment	12.99	2.22	0.31	0.70

Note : * In kilo metres; ** In metres
Source : Survey Data.

Government-initiated collective action is also found to concentrate on low-lying plots, which are near the outlet. However, the distances from the main canal and branch canals are comparatively high in respect of this institution. This pattern is more or less expected as government initiated collective action or the CAD Programme is intended for those plots, which are away from the canal flows but with possibilities of improved water availability, if seepage of water during transmission is effectively controlled.

The mean values of locational variables for farmer-initiated collective action indicates higher possibilities of the institution in those plots which are near to the branch canal and outlet, though the distance from the main canal is considerable. The average elevation of these plots from the outlet (though positive) is also low. All these indicate that, the institution is suitable in those plots, which have a possibility of improved supply of water if the farmers undertake some efforts.

Plots with collective action in lift irrigation as the institution are found located unfavourably in terms of distance from the outlet and the elevation of the plot. Average distance from the outlet is the highest for this institution. The fact that the plots are not distantly located from main and branch canals is indicative of the reliance of the institution on seepage water as the major source. Thus, it could be concluded that collective action in lift irrigation is resorted to in plots, which are clearly away from the outlet, and with considerable elevation.

Water market is found to be opted in those plots, which have the maximum elevation, and are at a considerable distance from the outlet. Due to the disadvantageous position, the possibility of water availability either through seepage or canal network is low in these plots. However, the distances from the main and branch canals are not that far indicating the necessary condition of a favourably located seller adjacent or near to these plots.

Private investment is concentrated in those plots, which are located at a distance from the main and branch canals, at a fairly high elevation. However, the plots are not so distant from the outlet, indicating the possibility of collecting water through seepage if wells are dug. High elevation and the distance from the canal flows are the main reasons that make private investment as the possible option in assuring irrigation.

Thus, from the analysis of the average values of locational variables, it is clear that the possibility of canal and canal based institutions is high in low lying plots while alternative institutions are more prominent in cases where plots have higher elevation. Also, the chances for alternative institutions increase with increased distances from the water flows in the main and branch canals.

Locational Concentration of Institutions in Paddy Plots

The mean values of locational variables for the 124 paddy plots shows that (Table 6.4) public provision in paddy is concentrated in those plots, which are located favourably in terms of distance from the water sources and elevation. In these plots, the assured supply of water obviates the need to go for alternative institutions. Government initiated and farmer initiated collective action are also found in plots with a similar elevation. However, in the case of government initiated collective action, the average distance from the main canal is found high compared to canal and farmer initiated collective action, in line with the policy priority under the CAD Programme. As far as farmer initiated collective action is concerned, the distance from the outlet is found to be higher which indicates the need for some effort on the part of the farmers to improve irrigation supply.

Table 6.4 : Mean Values of Locational Variables across Institutions for Paddy Plots

Institution	*DISTM**	*DISTB**	*DISTO**	*ELEV***
Public Provision	15.18	1.10	0.35	–2.26
Govt. Initiated Collective Action	19.60	1.65	0.34	–2.20
Farmer Initiated Collective Action	15.22	1.37	0.55	–2.41
Collective Action in Lift Irrigation	14.83	2.38	0.75	–0.50
Private Investment	12.89	2.25	0.54	0.55

Note : * In kilometres; ** In metres
Source : Survey Data.

Water market is absent in the case of paddy plots. The reason for this may be lower profitability of the crop that makes the option of water purchase non-viable. Further, as most of these plots are low lying, there are cost effective possibilities of improving the water supply with other alternative institutions.

Collective action in lift irrigation is found in those plots, which are visibly distant from the branch canal and the outlet. However, the negative elevation and nearness to the main canal confirms the possibility of seepage as a source of water for collective action in lift irrigation. In the case of private investment, the average elevation is found positive, unlike the other institutions. The distances from outlet and branch canals are also found high indicating the necessity for an alternative arrangement to assure water. The advantage of being near to the main canal assures seepage, which could be collected through wells.

Thus, the mean values of locational variables in paddy indicate a strong concentration of public provision and other canal-based institutions in favourably located plots. On the other hand, alternative institutions are found in plots with unfavourable locational variables.

The concentration of institutions in specific locational trajectories is indicative of the pattern of choice of the farmers, which reflects their cost considerations. Thus, the strategy of cost minimisation of farmers and transactions cost implications could indirectly be captured through locational variables. Locational attributes could affect cost in two ways. Firstly, these could influence the cost associated with the physical infrastructure requirement of various institutions (production costs). Secondly, location would also have direct implications on the cost associated with organising, enforcing and monitoring these institutions (transaction costs).

As has been clearly evident in the above discussion, public provision institutions are concentrated in plots, which are advantageously positioned in the canal command. The cost of physical infrastructure under public provision for all plots spread across the command area, irrespective of location, is marginal as these are set up by the state. Hence the only cost that is important in the context of institutional choice with respect to public provision is transaction costs. The fact that a farmer who is disadvantageously located does not opt for public provision thus indicates that the transaction costs associated with public provision in distant locations are higher. This is due to the positive relationship between water scarcity and the magnitude of locational variables. Increased water scarcity would also mean that farmers have to incur increased cost (time and effort) towards information gathering about the turns, negotiation with neighbouring farmers and canal officials, conflict resolution, influencing (both political and bureaucratic) and monitoring the implementation of turns. These higher transaction costs results in the choice of alternative institutions in disadvantaged plots.

Once the institution of public provision is found costlier, with higher transaction costs (due to the disadvantageous location of the plot) a rational farmer would select another institution, from all other alternative institutional options, which offers minimum total costs. The selection of different institutions in different plots indicates that the costs associated with these institutions differ across plots.

The production cost associated with collective action institution, does not vary much across plots with different locations as water is transferred mainly through the water-courses of the canal network. Here also, the only cost that could differ across farmers is the transaction costs. Thus, the concentration of plots under the collective action in comparatively better located (*vis-à-vis,* plots under other alternative institutions) suggests that the transaction costs of collective action are less in these plots. In other words, this reflects the lower resource requirement towards organising and maintaining the institution in these advantageously located plots.

As has been discussed earlier, all other institutions (such as collective action in lift irrigation, private investment and water markets) exist in plots, which are disadvantageously located in the canal command. Of these, water markets and collective lift irrigation exist in plots, which are located at highly disadvantageous positions, compared to private investment. Unlike public provision and collective action in canal, alternative institutions do incur substantial production costs. However, production cost associated with these institutions do not vary considerably across plots, and hence it is reasonable to assume that the selection of varied institutions in different plots are due to differences in transaction costs arising out of locational constraints. All these aspects are captured through some case studies in the subsequent session.

Cost Minimisation Strategy of Farmers : Evidence from Empirical Cases

Two sets of empirical cases are examined for understanding the cost minimising behaviour of farmers under various locational constraints. In the first set of cases—7.I (a) and 7.I (b)—an attempt is made to analyse the differences in institutional choices of the same farmer under varied plot characteristics. In the second set of cases, simultaneous existence of different institutions in the same area but with different locational variables is examined (case 7.II).

CASE 6.I : INSTITUTIONAL CHOICE OF FARMER WITH MULTIPLE PLOTS

CASE 6.I.1 : THE CASE OF A FARMER IN KARUKUTTY VILLAGE

A farmer in Karukutty village cultivates 4 plots of land under various institutions. The details of the plots are given below :

Plot 1 : In plot 1, for the last 13 years, the farmer is cultivating plantain and nutmeg. The source of irrigation is an open well constructed in the plot, where a 3 HP motor is installed. Before the construction of well and installation of the pump-set, the farmer was relying on public provision for irrigating the crops. Since the plot is located at the head reach (1.5 km. from Kizhakkumuri branch and 15 metres from the outlet) water scarcity is not severe and the plot had the advantage of seepage from the canal. The initial investment on well and pump-set was about Rs. 8500 (besides the subsidy of about Rs. 5600). The recurring cost, mainly the electricity charges, amounts to Rs. 320 per annum. For the farmer, continuing with public provision would not have implied any additional production costs, except that towards the maintenance of individual channels. In this context, how can the choice of the farmer for private investment be rationalized? A comparison of the costs associated with both the institutions is attempted which would provide some insights in addressing the above question.

Table 6.5 : Plot Details—Case 6.I (a)

Plot No.	Area (In cents)	Crops Grown*	Institution Followed	Previous Institution(s)	Locational Variables: Distance from Main Canal (km)	Distance from Branch Canal (km)	Distance from the Outlet (km)	Elevation of the Plot (metres)
1	170	Plantain, nutmeg and coconut	Private investment (Well)	Public Provision	10.50	1.5	0.15	1.52
2	35	Plantain, Coconut	Collective action in lift irrigation	Public provision	12.00	2.0	0.90	2.00
3	68	Plantain	Public provision	—	11.25	2.5	0.30	−2.25
4	212	Coconut	Private investment (Well)	Public provision	13.00	2.3	0.005	−3

Note : * Crops are indicated according to decreasing order of prominence in the plot.
Source : Survey Data.

Public provision involves zero one-time cost (both production and transaction). The recurrent costs associated with public provision are the expenses on the annual maintenance of field channels (production costs); and the transaction costs associated, such as information gathering about the turns, negotiation with neighbouring farmers and canal officials, conflict resolution, bribing of irrigation officials, influencing (both political and bureaucratic) and monitoring the implementation of turns. As against this, private investment implies considerable one time costs : expenses on constructing well and installing the pump-set (production costs) and those on gathering information on possibilities, organising loans and subsidies, efforts towards installing the system (transaction costs). The recurrent costs in private investment implies fuel charges and maintenance expenses (production costs) and negligible or almost zero transaction costs.

In public provision, though the plot was located near to the branch canal and outlet, assuring an adequate level of supply implied considerable resources in acquiring water, due to the position of the plot at a higher elevation. For instance, the farmer reported that in an year on an average 8-10 days manual effort (either hired or own) is required to divert water to the plot. Apart from this, substantial time and efforts are required towards negotiation and conflict resolution with other irrigators and the authorities. Given this, a shift towards private investment offers considerable scope for minimising the costs of acquiring water, with significant reductions in recurring transaction costs. Further, the farmer also perceives growing scarcity of canal water in the future, which will increase the transaction cost of public provision in terms of time and effort required to assure irrigation. The farmer assigns a higher cost on the time spent on irrigation as the location of the plot is away from the place of residence of the farmer. Thus, in the long run the transaction cost of public provision (mostly time and effort) to ensure adequate supply to meet the requirements is perceived to be higher than the recurring cost on the well. The larger recurring cost associated with collective action, especially in terms of time and effort is the major factor that was highlighted to restrict collective action as a feasible option.

The farmers also expect low profitability of crops with canal irrigation on account of the difficulty in farm planning due to uncertain water availability. The possibilities of water purchase and collective action in lift irrigation also do not come under the farmers' consideration, as these are costlier to organise.

Plot 2 : Collective action in lift irrigation has been the institution in plot 2 for the last 18 years. The elevation of the plot is 2 metres above the outlet level and is distant from the outlet (0.9 km), and water scarcity has been a problem from the time of commencement of the canal system in the locality. This induced the farmer to pressurise the authorities, along with other similarly placed farmers, to exempt the plot from the cess. After obtaining the exemption, a group of 8 big farmers and 6 small farmers established a lift irrigation system, which pumps water from the Chalakkudy River with a 10 HP motor (which was bought with a subsidy of 40 per cent from the Agriculture Department). At the time of joining, the farmer had incurred an initial fixed cost of Rs. 500. The present annual recurring cost ranges between Rs. 100-150 per acre.

Here also, the production costs associated with public provision are the expenses on the annual maintenance of field channels (production costs). Apart from production costs, costs are involved on information gathering about the turns, negotiation with neighbouring farmers and canal officials, conflict resolution, bribing of irrigation officials, influencing (both political and bureaucratic), monitoring and enforcement of the turns. Under collective action in lift irrigation, the costs of irrigation involve farmers' contribution towards the construction of facilities and share of costs incurred on information gathering, consensus-building, negotiation and organisation. The costs also include water charges, individual share in the cost of maintenance and expenses on conflict resolution, monitoring externalities and enforcement.

Due to the disadvantageous location of the plot, the appropriation of adequate water through the canal system implies tremendous costs on information gathering, negotiating and conflict resolution, to ensure a desired level of water supply. All this would require several days of effort of the farmer, apart from engaging hired workers for few days to divert water to the plots. Due to acute scarcity of canal water, collective action institutions are also costlier on account of high costs associated with organising, monitoring and enforcing the institutions. Private investment would also mean costlier due to poor ground water availability and low seepage from canal. Though possibility of buying water is open to the farmer, water charges and the costs of negotiating and enforcing water purchase annually would entail substantial costs, which makes water purchase a costly proposition compared to collective action in lift irrigation.

Plot 3 : In Plot 3, farmer continues with public provision. The plot has adequate water supply as it is below the outlet level and near to the outlet. Apart from the manual effort, which is negligible, no costs are involved and hence the farmer does not have any incentive to look for any alternative institutions that are obviously costlier, compared to the present institution.

Plot 4 : Private investment is the institution in Plot 4. The initial cost of setting up the well and pump-set is Rs. 35,000. Though the plot is located very near the outlet (5 metres), its elevation (3 metres) from the outlet level, constrain the easy access to adequate water. To ensure adequate irrigation supply, the farmer would incur significant costs which include the time and effort in negotiating with the neighbours, monitoring and enforcing the turns and resolving conflicts, which would have induced the farmer to invest on well and pump-set. As the outlet is near to the plot, well assures adequate water and reduces the time required in acquiring water. Thus, in the long run, this could be viewed as one with the lowest transaction cost. Investment on well and pump-set lead to fixed timing in irrigation, thereby lessening the time spent on irrigation. Organisation of collective action institutions is also costlier since the area does not have general water scarcity. Further, institutions such as water markets and collective action in lift irrigation are also costlier as the organisation of these would entail substantial costs on organizing, negotiating, enforcing and monitoring compared to well.

CASE 6.I.2 : THE CASE OF A FARMER IN MELUR VILLAGE

A farmer in Melur cultivates plantain in three plots. The summary details of the plots are as shown in the Table 6.6.

Plot 1 : Plot 1 lies near the branch canal (1 km). However, the plot is distant from the outlet (0.80 km) and has an elevation of 0.50 metres. The proximity to branch canal assure availability of seepage which prompted the farmer to dig well and install pump set (1.5 HP). The total cost on this was about Rs.6,000, apart from a subsidy of Rs. 1200 from the Agriculture Department. The current cost of acquiring water is equivalent to the electricity charges, which comes around Rs. 150-200.

Ensuring adequate water through public provision requires considerable efforts towards negotiating with fellow irrigators and enforcing the turn system. Collective actions are also perceived costlier, considering the time and effort required to assure adequate irrigation. The major advantage of opting for well is the reduction in cost in terms of time and effort by way of negotiating and monitoring.

Plot 2 : The plot is located at a higher plain from the outlet level (1.50 metres) and is also distant from the outlet. Irrigation requirement is met through purchase of water (at the rate of Rs. 6 per plantain) from an adjacent and favourably located plot. Public provision and collective action institutions demand considerable time and effort per unit of irrigation, in the context of the elevation of the plot. The farmer had some previous experience with farmer initiated collective action. The shift to water market was on account of the higher cost in collective action in terms of time and effort to ensure the adequate supply. Further, as the plot is leased in, the farmer does not have any long term considerations on the plot, which prevents from investing on well or trying for collective action in lift irrigation, which imply substantial one time investment.

Plot 3 : Collective action in lift irrigation is the institution for water provision in Plot 3. The plot is distant from the branch canal (4.5 km) and is above the outlet level (1.00 metre). The rent inclusive of water charges is Rs. 8 per plant and there are around 500 plants in the plot. The reason for going for collective

Table 6.6 : Plot Details—Case 6.I (b)

Plot No.	Area (in cents)	Crops Grown*	Current Institution	Previous Institution(s)	Locational Variables			
					Distance from Main Canal (km)	Distance from Branch Canal (km)	Distance from the Outlet (km)	Elevation of the Plot (metres)
1.	30	Plantain	Private investment (Well)	Public provision	12.30	1.00	0.80	0.50
2.	50	Plantain, tapioca	Water Market	Public provision	11.50	2.00	0.75	1.50
3.	100	Plantain	Collective action in lift irrigation	Public provision, Govt. initiated collective action, Farmer initiated collective action	9.30	4.50	0.60	1.00

Note : * Crops are indicated according to decreasing order of prominence in the plot.
Source : Survey Data.

action in lift irrigation is due to the failed canal and collective action initiatives. Since the area has acute water shortage, the perceived costs of acquiring water through public provision or collective action institutions are high compared to that of collective action in lift irrigation. To ensure adequate irrigation water, negotiations and bargaining are required not only with neighbouring farmers but also with farmers in the middle and head ends as well as the concerned irrigation authorities. Since, acute shortage exists, with supply always falling short of demand the possibility of violation of schedules and turns are found high which increases conflicts and hence collective action institutions have failed. Private investment is also not considered as a viable option as water availability through seepage is low. Water purchase, is also a costly option for the farmer, given the higher water rates and organising costs associated with it.

CASE 6.II : MULTIPLE INSTITUTIONS IN THE SAME LOCALITY—THE CASE OF MANJAPRA

In Manjapra village, 7 plots, which are almost uniform in terms of distances from main and branch canals are found having different institutions. The plots are all located around 20 kilometres from the main canal and 2.50 to 3 kilometres from the branch canal. The plot-wise details are given in the Table 6.7.

CASE 6.II.1 : THE CASES OF WELL IRRIGATION

Well irrigation was found in three plots. All these plots are situated near to the outlet and hence the availability of water is comparatively high. However, all these plots are situated at a higher elevation. The availability of seepage in the plots is the main reason for investing on wells. The farmers' option for well was sequential to the failed attempts under previous institutions (public provision and collective action) in ensuring adequate water. To assure the same level of water availability through public provision or collective action institution, the

Table 6.7 : Plot Details—Case 6.II

Characteristics	*Plot Number*						
	1	*2*	*3*	*4*	*5*	*6*	*7*
Institution	Private investment (Well)	Private investment (Well)	Public provision	Water market	Water market	Private investment (Well)	Public provision
Previous Institutions	Public provision, Farmer initiated collective action	Public provision, Farmer initiated collective action	Public provision, Farmer initiated collective action	Public provision, Farmer initiated collective action	Public provision, Farmer initiated collective action	Public provision, Farmer initiated collective action	Public provision, Farmer initiated collective action
Crops Grown*	Coconut, plantain	Coconut, arecanut, nutmeg	Paddy	Plantain	Coconut, plantain	Coconut, plantain	Paddy
Land-size (In cents)	200	155	80	100	80	70	60
Distance from the outlet (km)	0.20	0.05	0.75	0.90	1.00	0.30	0.60
Elevation of the plot (metre)	1.00	0.75	–2.00	1.50	1.50	0.50	–1.5

Note : * Crops are indicated according to decreasing order of prominence in the plot.
Source : Survey Data.

farmers perceive higher costs, especially in terms of wages of hired workers and own efforts. The other options of collective action in lift irrigation and water markets are still costlier as the farmer has to spend excessively on organising the institution.

CASE 6.II.2 : THE CASES OF PUBLIC PROVISION

Both the plots under public provision are situated below the outlet level, though relatively far-off from the outlet. Paddy is the main crop grown in these plots. The plots are left fallow during *Puncha* season. Investing on well and pump-set is considered less rewarding, given paddy as the main crop. Similarly, the institutions of collective action in lift irrigation and water markets are also perceived costlier compared to the benefits from paddy cultivation. With many farmers in the locality going for alternative institutions, especially wells, the cost of organising collective action among farmers is also perceived to be high.

CASE 6.II.3 : THE CASES OF WATER MARKETS

The institution of water market is found in two plots. The plots are away from the outlet and are located above the outlet level. The relatively high cost on ensuring adequate water, especially cost of organising and monitoring, through public provision and collective action institutions have resulted in the shifting to water market. Due to locational disadvantage of the plots in terms of distance from the outlet, investment on well and pump set is seen as costlier compared to the present institution due to the large fixed investment required. Collective action in lift irrigation would also mean high costs in organising, as the institution is absent at present in the vicinity.

The empirical cases confirm that the net pay-off maximisation of the individual farmer is that of a cost minimisation problem, constrained by locational variables where transaction cost assumes a central concern. A rational farmer's choice of an institution in any particular situation depends on how the individual perceives transactions costs associated with various alternative institutions, given the production costs. The fact that the same farmers opt for different institutions in different plots highlights the importance of locational factors in the cost calculations of various institutional choices. The production cost associated with the same institution does not vary much across plots. Thus, a farmer's selection of multiple institutions in different plots indicates the variations in transaction cost associated with the same institution across plots. The case studies, thus show how farmers under different locational constraints select institutions that minimises transaction costs (especially *ex post*) of irrigation, even in situations where production costs are significantly higher, thereby minimising individual's total costs. The cost minimisation strategies of individuals, underlying the institutional choice are primarily aimed at attaining individual efficiency. Nevertheless, the efficiency and aspects of the individual cost minimisation needs to be understood from the social point of view, to assess the overall efficiency and equity aspects of these institutions, which is addressed in the subsequent chapter.

7

Impact of Irrigation Institutions on Efficiency and Equity

From the previous chapters, it is evident that multiple institutions exist in the canal command of CRDS and that individual farmers' choice of institutions are based on the cost minimisation strategy within a set of constraints. It is important to analyse whether these individual choices would lead to a socially efficient allocation of resources. This is important due to the existence of certain features associated with canal irrigation, such as public good, externality and economies of scale that would lead to a gap between private and social gains. The implications of the existence of multiple institutions for water distribution on social efficiency are the major issues discussed in this chapter. Further, it also discusses the equity implications of these institutional choices.

Canal irrigation is often characterised as a public good with unique socio-economic, institutional and engineering aspects. Firstly, it is often argued that it is difficult to leave development and allocation of surface irrigation water to free market, on account of problems in defining private property rights due to issues of costly exclusion and subtractability. Secondly, centralised coordination is considered superior due to natural monopoly and the related scope of significant economies of scale that can be gained in the storage, conveyance and distribution of large quantities of irrigation water. Further, centralised agency is assumed to deal with externalities efficiently, arising out of development and management of water sources. Finally, and the most important of all, is the attainment of social objectives such as income redistribution, food self-sufficiency and sustainable agricultural production. The first three sets of issues relate to the efficiency, while the last set is concerned with equity. In the subsequent sessions, these issues are analysed in the context of irrigation institutions in CRDS.

Public Good Character of Canal Irrigation

The efficiency rule for a public good[1] is that the marginal cost to the society of providing the good should equal the sum of the marginal benefits for all consumers—the marginal social benefit. The marginal social benefit function is derived through the vertical summation of the marginal benefits (or willingness to pay) of all the individual consumers. In the case of a pure public good, the services are available to everybody, irrespective of their individual demands. Furthermore, it is prohibitively costly to exclude non-payers from using the service. Irrigation also has some nature of a public good as it is costlier to exclude non-payers.

Thus, institutional choice in canal command can lead to overall increase in social efficiency only if the level of provision is up to the point where marginal costs equals the sum of the marginal willingness to pay of all the users. But in institutions such as farmer initiated collective action, there are possibilities of non-cooperation and free riding by some of the beneficiaries. Anticipating this possibility, those who cooperate would contribute towards the institution only upto the point, where marginal costs equals the sum of their marginal willingness to pay (and not that of all users). This is a source of social inefficiency. Various cases from the CRDS command area provides empirical evidences to indicate this type of inefficiency.

CASE 7. I : A CASE OF FAILED CO-OPERATION IN MAMBRA

In Mambra village, collective action of farmers for cleaning the branch canal and water-courses was successfully organised, till 5 years ago. The cost of these activities was shared by the farmers according to the size of their plots, and the farmers recollect that the cost associated with this institution was limited to a few days' effort and some minimal monetary contribution. Of late, some of the farmers, who are located favourably, started keeping away from collective action, as water availability in their plots is assured, irrespective of their participation in collective action. This withdrawal resulted in the failure of the institution as other farmers found it costlier, to charge the free riders, in the absence of any mechanism. Subsequently, most of the farmers started investing on open wells and pump-sets. Here, the institutional change represents a case of social inefficiency as all these farmers have had to incur substantial expenses in setting up and maintaining wells and pump-sets, which is greater than what they would have incurred for a well functioning collective action at the local level.

CASE 7.II : A CASE OF FREE-RIDING IN EDAKKUNNI

In Edakkunni, due to inadequate water supply from the canal, farmers in the locality collectively established a lift irrigation scheme, where water was pumped from a village pond using a 30 HP pump-set. Prior to the formation of this scheme, all the plots had acute ground water problems. However, with the effective functioning of the lift irrigation scheme, water availability (seepage) in many plots improved considerably. This prompted some of the farmers (6 out of 22) to discontinue from the community irrigation scheme and to invest on wells and pump-sets, as they found it cheaper in terms of perceived reduction in time and effort to irrigate the plots. This withdrawal had increased the per acre monthly contribution of other farmers from Rs. 60-70 to Rs.80-90. The shift in the choice of few farmers in this case, can lead to a deterioration in social efficiency, if the aggregate gain made by the farmers who have shifted is less than the aggregate loss incurred by those who continued in the lift irrigation scheme. This seems to be the case in Edakkunni.

Public Provision as a Natural Monopoly

Irrigation due to its natural monopoly character, have potential for economies of scale. It is likely that, the average cost of irrigation can be reduced with larger scales of operation. The economies of scale, here, might be operating at different levels. At one level, there are economies of scale in the whole system of storing, conveying and distributing water, and thus larger systems may lead to a lower per unit costs than that of a number of small systems as a whole. Secondly, there may be economies of scale in water distribution at the field level. There may be cost advantages if a large number of farmers in a locality use the same institution of water distribution. The inability of these farmers to come together may lead to evolution of a number of institutions and this may increase the cost as a whole. This could be viewed as yet another source of social inefficiency.

Viewed in this perspective, if the sum of the additional cost incurred by all the farmers exceed the cost at which the state or any other institution would have provided certain and adequate supply of water, then the existing institutions could be seen inefficient. In the case of CRDS, if the gross additional cost incurred by farmers on different institutions would have been collected by an agency or government and channelised for the maintenance of the system, and if it results in increased supply and certainty, this implies that the *status quo*, characterised by a poorly functioning centralised system and a number of micro institutions of water distribution is socially inefficient. It is observed in many plots in the mid reaches of the command area that farmers invest on open wells and pump sets to counter the uncertainties arising from irregular water turns. Here, the emergence of private investment could be viewed as a socially inefficient outcome, as the possible costs of streamlining the turn system in the public provision would have been far less, compared to the aggregate of the individual expenses on private investment. The following case provides insights into this aspect of social inefficiency of institutional choice.

CASE 7.III : DISECONOMIES OF PRIVATE INVESTMENT IN AZHAKAM

Azhakam branch canal is located in head end of the Kalady Main Canal and hence water availability is fairly good in the canal. However, it is observed that, almost all the farmers located in the middle reaches of the branch canal have invested on wells and pump-sets. This shift towards private investment was for minimising the efforts on diverting the irregular and untimely irrigation turns to the individual plots under public provision. In this case, the total costs (summation of the costs of individual farmers) could have been reduced considerably if a centralised agency or a farmer collective had organised a mechanism to design and enforce a definite turn system. Thus, the institutional change towards private investment could be viewed as socially inefficient. So in this case, even though the private investment turns out to be the cost minimising solution for the farmers considering the constraints of water availability in public provision, and the high transaction costs in mobilising community action, such an alternative exercised by a large number of farmers leads to socially inefficient allocation of resources.

However, the use of institutions alternative to public provision need not always produce socially inefficient outcomes. For instance, there could also be situations where individual or group actions by the farmers would result only in a lower aggregate (or equal) cost compared to the one required to provide the same service through public provision. During the field survey, it was noticed that some of the disadvantageously placed farmers were jointly lifting water from ponds or canals on a cost sharing basis or purchasing water. In some of these cases, even if a central agency undertakes to provide irrigation, the possible expense would have been more or less the same, which would mean that the individual choices of alternative institutions are also socially efficient. The following case studies give insights into this aspect.

CASE : 7.IV : COLLECTIVE ACTION IN LIFT IRRIGATION IN NADUTHURUTHU

In Naduthuruthu, following acute water shortage under public provision, a group of farmers organized a community lift irrigation scheme, under which farmers constructed a check-dam across a natural stream and diverted water to a village pond. The water from the pond is then lifted with electric motor and distributed through field channels. The scheme covers 24 acres of land belonging to 45 households. All these plots are situated at the tail end of a branch canal and are at higher elevations compared to the outlet. The establishment cost of this scheme was about Rs. 40,000, which was raised from among the beneficiaries. The scheme is running successfully since 1976. The recurring cost is shared between the beneficiaries on a per acre basis, which is Rs. 40 per month. In this case, the institutional change towards collective action in lift irrigation can be viewed as socially efficient, as the cost of providing the same level of supply through public provision or any other institution would entail a higher cost in the absence of participation of irrigators.

CASE 7.V : COST EFFECTIVENESS OF WATER MARKET IN KALADY

In the tail end of Kalady Branch Canal, three farmers were found resorting to purchase of water from a neighbouring farmer, who lifts water from a well using a 3 HP electric motor and charges Rs. 9 per hour. The favourable positioning of the seller enables him to collect the seepage water from the canal and sell the excess water (after meeting his own demand) to the nearby plots. In this particular case all the buyers are disadvantageously placed at higher elevations or distant location from the water sources. As lifting water is the only way to irrigate these plots, and since there is no scope for scale economies any other institution would entail a much higher cost than the *status quo* institution to ensure the same quantum of water supply and hence the institution could be viewed as socially efficient.

An institutional choice in irrigation could lead to social inefficiency, in cases where individual or small groups of farmers have to negotiate and resolve conflicts on their own. There may be economies of scale in carrying out such negotiations through a larger body of irrigators. Moreover, the costs of conflict resolution would be significantly lower if a centralised agency accountable to a larger body of water users can sort out the conflicts between different types of users. For instance, for a tail end farmer or a group of them, it may involve very high costs or in some cases not practicable, to sue the head-enders who are over irrigating or misusing the irrigation water. These issues could be resolved at a much lower per capita expenditure, with the help of a well functioning centralised agency, accountable to all the users.

Externality in Irrigation

Irrigation is also subject to issues of externalities. Negative externalities in irrigation arise when the institutional choice of certain irrigators lead to reduced availability of water to other farmers who are also entitled to use the resource. This issue is closely related to the subtractable nature of irrigation water.

In the case of canal irrigation, unlike other economic goods where scarcity prompts efficiency and conservation, water is liable to be misused. Farmer's right to quantum of irrigation water is not well defined and inadequately enforced. This, coupled with the absence of volumetric pricing, lead to a situation where each farmer can use as much water as possible affecting the supply available to others. Overuse of irrigation water by farmers near the head of the system often leads to the deprivation of water to the tail-end users. For instance, during the survey, many farmers in the head and middle ends reported that they flood their plots whenever water is available in the canal. This is mainly due to the perceived uncertainty in the supply through public provision. This over-use in turn results in reduced availability of irrigation at the tail ends, which necessitated the emergence of alternative institutions especially private investment. These are indicative of a case of social inefficiency arising out of negative externality.

Apart from this over use, appropriation of canal water (directly or indirectly) through various institutions located in the head reaches or other advantageously located positions, may lead to a decreased water flow in the canal which ultimately reduces the availability of irrigation water to other farmers and thus would affect the functioning of existing institutions at the tail-end. Most of the alternative institutions, especially private investments, have shown concentration at specific locations in the head and middle reaches of the canal command where canal water is relatively abundant *vis-à-vis*, the tail-ends. Alternative institutions, while improving the water availability at these locations would reduce the flow to downstreams. The proliferation of open wells and in the head and middle reaches of CRDS is reported to have adversely affected the flows in the lower reaches of the command. In some other cases, farmers who have invested on wells and pump-sets reported that the lining of field channels through the CAD Programme has resulted in poor water availability in their wells. Institutions such as water markets do have substantial externality implications. In CRDS, there have been many reported cases of reduced water availability due to the extensive operation of water markets. The following case studies further highlight these aspects.

CASE 7.VI : EXTERNALITIES OF PRIVATE INVESTMENT IN BHOOTHAMKUTTY

Bhootamkutty is the second branch canal in the left bank canal network. Scarcity in water supply was not a major problem, even in the plots at the tail-ends, till recently. However, with the emergence of several open wells in the plots at the head and middle reaches of the branch canal, water shortage has cropped up as an issue among the tail-end farmers. The farmers at the tail-end perceive that the shortage is not due to the reduced release of water to the branch canal, but because of the massive transmission loss due to the digging of a large number of wells by advantageously located farmers. To counter this shortage, many of the tail-enders have also now invested on wells (which are deeper than those at the head reaches). All these are indicative of a situation of deterioration in social efficiency, arising out of externality issues.

CASE 7.VII : EXTERNALITIES OF WATER MARKET IN KUNNAPPALLY

In Kunnappally, scarcity of irrigation water under public provision became acute since one of the advantageously placed farmers in the area started selling water from wells located in his plot to some of the adjacent plots using two 5 HP electric motors. To restrict this negative externality, a group of 37 farmers filed a court case against the seller. It was reported that, the over extraction of water (seepage) by the seller reduces the water available to the plots in the lower reaches. In this case, the cost of negotiating and resolving the issues (transaction costs) arising out of negative externalities could be seen as the major form of social inefficiency.

Other Efficiency Aspects

Every farmer who has opted for alternative institutions had something to gain, especially in terms of certainty and adequacy of water. All these institutions can be seen as arrangements to do away with the confusions about entitlements and its enforcement. The emergence of alternative institutions is driven by the recognised need for order and stability in the expectations and behaviour of farmers. Every institutional change, thus, adds to the certainty and adequacy of irrigation. Further, given the poor state of centralised irrigation system and the inabilities to achieve enhanced efficiency within that system, an improvement in use efficiency of the existing irrigation potential realised through new institutions, would imply an increase in efficiency. Thus, it can be viewed that, new institutions improve water-use efficiency, since these are resorted to when the existing institutional framework fails to solve the fundamental problem governing water-use.

Unlike canal irrigation, the possibility of overuse and misuse of water tend to be less under alternative institutions, due to the prevalence of substantial positive costs. Thus, the extent of inefficiency in irrigation water use in other institutions will be less compared to canal irrigation. In CRDS canal command, this is particularly true with institutions such as private investment and water market as every additional pumping means increase in cost. The argument that water markets promote investment in increasing the efficiency of water use and to transform water from being a scarce but free resource into an economic good with an opportunity cost is thus true in this context.

Water markets tend to induce farmers to reduce irrigation frequency by best utilising the available soil moisture from precipitation and previous water application. Further, institutions like wells and water markets allows conjunctive use of surface and ground/seepage water in the canal command. With the development of water markets, upstream farmers are also found economising the use of irrigation water for selling the excess water to other farmers, who pay price, which is higher than the marginal value of water application in the sellers' plot. Water markets also help to earn a return on irrigation investment and leads to the fuller utilisation of irrigation assets such as wells and pump-sets and encourage new investments in area with adequate water resource potential.

Economic efficiency of new institutions is perceived to be higher, if irrigation water is diverted for high value use. Viewed in this perspective, in CRDS, the situation is certainly that of an improved aggregate economic efficiency, as most of the alternative institutions are found catering to non-paddy crops with higher profitability. This is evident from the approach of the farmers, as has been seen in the previous chapters. Most of the farmers who incur substantial additional costs in alternative institutions like private investments and water markets are found cultivating cash crops such as coconut, plantain, arecanut or nutmeg. Another argument in favour of increased economic efficiency is that almost all the alternative institutions harvest the otherwise wasted seepage water and divert it for value addition. For instance, most of the lift irrigation projects and water markets are based on seepage water, which would have otherwise gone untapped.

Equity Considerations of Institutional Change

The increased social efficiency of alternative institutions does not necessarily mean that the gains from the institutional change are equitably distributed. Equity considerations are important especially in a canal command, as all the farmers are legally entitled to get irrigation water. Equity in distribution in its complete sense is a non-existent proposition given the locational disadvantage of tail-enders *vis-à-vis* the head-end farmers. Thus, the concept of equity needs to be understood as some degree of fairness in the use of water rather than equal sharing by all the irrigators.

Equity considerations involve rights, access to, control over and distribution of irrigation. Issues of equity arise when a farmer appropriates excess irrigation water through any institution at the cost of other potential users who also have equal right to canal water. Appropriation of canal water (directly or indirectly) through various institutions located in the head reaches or other advantageously located positions, may lead to a decreased water flow in the canal, which ultimately reduces the availability of irrigation water to other farmers.

The case studies discussed above in the context of externality (Cases 7.VI and 7.VII) clearly involve equity issues. As seen in the case of Bhoothamkutty (Case 7.VI), the tail-enders have to incur a higher cost in acquiring water, while the farmers at the head reaches have over irrigated, without incurring any additional costs. In the case of water market in Kunnappally (Case 7.VII), 37 farmers are affected by the over appropriation of one single farmer, indicating a strong case of inequity associated with water market institutions. Here, the seller makes substantial gains from the appropriation of canal water, which would have otherwise been available to others. The adversely affected farmers have to bear not only the cost of crop loss in the absence of adequate irrigation but also the additional cost on conflict resolution. Several other cases of conflict on the issue of equity and property rights of water in the context of water sales have been observed in the canal command, during the survey.

In the case of water markets, water is essentially being privatised (private appropriation of public resource). This obviously leads to unequal distribution of benefits in favour of few farmers and these institutions lead to an overall reduction in the equitable availability of resources. The glaring inequity between unequally placed farmers, in terms of appropriation of benefits or sharing of burden, it is evident from a comparison of a water seller and a tail-end farmer. While the seller appropriates a large quantity of water (much more than the irrigation requirement of the plot) at a nominal cess and earns income through rent seeking, the tail-ender is denied of irrigation water, and above all is liable to pay water cess.

Water markets also brings into the forefront the issue of over-exploitation of buyers. Water is mostly transferred through unlined channels and the extent of seepage loss is high resulting in paying for unused water by the buyers. They are generally interested in getting water from the closest seller so as to minimise the cost involved. This would give rise to seller's monopoly over buyers located in the neighbourhood. Water markets, thus, have strong negative equity aspects with the possibility of rent seeking and skewed appropriation of gains by advantageously located farmers, at the cost of others given the lack of well-defined property rights on irrigation water.

The equity issues also arise, in cases of free rider problems, when the benefits arising from an institutional change is appropriated by those who do not contribute to the cost of the institution. Such instances are common in institutions of collective action (such as cleaning of water courses, lining of field channels and so on) where the benefits are shared even by those who have not contributed towards the institution. There are many cases where the water supply at the middle ends got improved due to the efforts of farmers at the tail-end, without any cooperation by the farmers in the mid reaches. Here, often the benefits to non-contributors (middle end farmers) were high compared to those who cooperated in the collective action, leading to inequity issues.

Notwithstanding the above aspects of inequity issues associated with institutional choice, it is erroneous to conclude that all the alternative institutions would necessarily lead to inequitable outcomes. There could also be cases where, the choice of alternative institutions leading to more equitable outcomes. For instance, institutions such as collective action in lift irrigation and water market, which are mostly concentrated in disadvantageously placed locations, have positive equity impacts, as it provides irrigation water to plots which have acute water shortage and hence with lesser scope for other institutions. One positive equity implication attached to water market is that it provides irrigation facility to those farmers who could not afford to make large investments in terms of wells and pump-sets or farmers who are at a locationally disadvantageous position.

The foregoing discussion clearly brings in the need to address the social efficiency and equity aspects of institutional choice of individuals. The institutional choice of farmers could result in increased social efficiency if the gainers of change can compensate the losers and still have some positive benefits. However, in the absence of clearly defined property rights, entitlements and enforcing mechanisms the scope for compensatory provisions does not exist. In reality, the losers are also liable to pay the cess, irrespective of the water availability, leading to a situation of worsened equity. For instance, several court cases have been pending on the issue of property rights and entitlements, without any definite decisions in the absence of a clear-cut definition of water rights. These ambiguities could be minimised through establishing and enforcing property rights in irrigation, which would eventually lead to socially efficient and equitable outcomes.

NOTE

1. A pure public good is a good or service, which is consumed by everyone and from which no one can be excluded. It has two characteristics, non-rivalry (non-subtractability) i.e. one person's consumption of the good does not reduce the amount available for someone else and non-excludability i.e. no one can be excluded from the consumption of the good (Foldvary, Fred; 1994).

8

Summary and Policy Implications

Irrigation policies in developing countries are facing daunting challenges with the weakening of the emphasis on state ownership and management. The dismal performance of the state-run irrigation, in terms of low recovery of the operation and maintenance expenses, poor quality of service and the mounting costs of developing new sources of water have evoked considerable attention and concern leading to new initiatives and efforts to explore institutional alternatives. The three policy prescriptions often cited in the context of the failure of public provision are : reforms of public management of irrigation systems; community management of irrigation and the establishment of private property rights.

In India, in pursuance of the Government's irrigation policy, considerable investments have been made by the state on the development and maintenance of large scale irrigation projects with a view to enhance agricultural production. However, the reports of the poor performance of large scale irrigation projects, set up at huge costs, and their consequent inability to realise the envisaged objectives have underscored the need for institutional reforms in the management of canal irrigation. Notwithstanding the extensive debates on finding alternative institutional solutions, the reforms introduced have been largely confined to the transfer of irrigation management to user groups ignoring the existence and possibilities of other institutions. In reality, a wide range of institutions are involved in the provision of irrigation. A closer examination of various institutional alternatives and their functioning can enrich the understanding of institutional choice and hence could contribute towards framing of suitable policies to improve irrigation management. With these objectives, the study examined the emergence and functioning of multiple institutions and their efficiency and equity impacts in the canal command of Chalakudy River Diversion Scheme (CRDS), the first major surface irrigation project in Kerala.

Summary of Findings

The most important reason attributed to the failure of surface irrigation projects in Kerala is the gross neglect of the specifics of the State's agriculture. The undue importance given to paddy cultivation, which has been the stated objective of the agricultural and irrigation policies of the State, is found to have invoked rigidities in the institutional framework of irrigation. The stress on large scale irrigation projects has also neglected the agro climatic specifics of the State, which has natural advantages for other cost effective irrigation possibilities. Administrative flaws in terms of improper management and low cost recovery have further accentuated the gravity of the issue. The growing mismatch between the provisions of canal irrigation and the specific needs and requirements of farmers at the micro-level, have brought in responses towards modifying/improving the canal system or in crafting of alternative institutions. To capture this a survey of 397 plots spread over various reaches of the canal command of CRDS was carried out.

The survey identified a host of factors which contributed towards the failure of canal system in CRDS, such as : inadequate operation and maintenance, unplanned extension of canal due to political considerations, administrative mismanagement, overuse and misuse of water by head-enders and changed irrigation requirement due to shift in cropping pattern in favour of non-paddy crops. These issues associated with public provision, are found to have resulted in farmer responses leading to the emergence of alternative institutions in the canal command for the distribution of water. These institutions reflect the choice domain of independent economic actors at a given point of time.

The choice of an individual farmer differs from that of others and can vary over time. However, at a given point of time, there may be some clear-cut patterns of choice, which may be governed by formal or informal rules in use. In the context of the present study, each set of formal or informal rules or norms, governing the modes of acquiring irrigation water is considered as an institution. Accordingly multiple institutions were found functioning in the command area of CRDS. These institutions are broadly categorized as : Government Initiated Collective Action; Farmer Initiated Collective Action; Collective Action in Lift Irrigation; Water Market; and Private Investment. Only 15 per cent of the plots studied were found depending solely on public provision, highlighting the prominence of alternative institutions in the canal command. Among the alternative institutions, private investment caters to the largest proportion of plots (43.3 per cent) followed by farmer initiated collective action (21 per cent) and water market (12.6 per cent).

Institutional choices in irrigation is analysed from the perspective of an individual farmer who is assumed to be maximising pay-offs. The individual's choices of institutions in any situation is taken as a function of expected benefits and expected costs, given other constraints. Given the existence of multiple institutions in irrigation, a farmer selects that institution which maximises the net pay-offs given the constraints. The costs-benefits calculations and hence the matrix of pay-offs of institutions differ across plots and individuals, which in turn depend on a host of factors. Three sets of factors are delineated as the major determinants in irrigation decisions of the farmers. These are : (a) locational factors; (b) land and crop specifications; and (c) individual and household characteristics. The nature and extent of the relationship between these factors and their influence in the institutional choice were analysed using a multinomial logit model. For estimating the multinomial logit model, the various institutions in the canal command are broadly categorised into three sets—*public provision; collective action in canal;* and *other institutions. Collective action in canal* includes both Government initiated and farmer initiated collective actions and *other institutions* include : collective action in lift irrigation, water markets and private investment.

The analysis of the effect of locational factors on institutional choice revealed that these factors are significant in the choice of both *collective action in canal* and *other institutions*. In the case of *collective action in canal*, the effect of distance from the main canal, branch canal and outlet are found significant and negative indicating the lower chances of opting the institution beyond 10 kilometres from the main canal, 2 kilometres from the branch canal and 0.5 kilometres from the outlet, respectively.

In the case of *other institutions*, prominence of significant effects is observed for all locational variables. Distance from the main canal, branch canal and elevation of the plot are found to be positively related indicating higher chances of *other institutions* (outside canal) in distant locations from the canal network, where water is scarce and uncertain. The direction of effect of distance from the outlet was found negative, which could be attributed to the lower likelihood of farmers incurring considerable investment, when the possibility of getting seepage water is unsure.

As regards the land and crop characteristics, the main crop cultivated is found significant in both *collective action in canal* and *other institutions*. In the case of *collective action (in canal)*, this is negatively related indicating lower chances of non-paddy growers opting for canal-based institutions than *public provision*. In the case of *other institutions*, non-paddy as main crop is found to have a positive relationship. This is indicative of the fact that farmers are ready to go for alternative irrigation institutions outside canal, if the main crop cultivated is not paddy. Land type is found significantly and negatively related in the case of wetland revealing lower chances of *other institutions* in wetland plots.

Among the household characteristics, education is found significant and positive in both the cases of *collective action in canal* and *other institutions*, which shows higher preferences for alternative institutions among educated farmers. Cultivation as main occupation is negatively and strongly significant in the choice of *other institutions,* which explains the strong preference among full time cultivators not to go for *other institutions*.

In the CRDS command area, the crops cultivated are broadly of two types; paddy and non-paddy. Non-paddy plots are mostly under mixed cropping with coconut as the main crop along with, nutmeg, plantain or arecanut as main subsidiary crop. The disaggregate analysis of the determinants for these two crop categories also identified locational variables as the important determinant of institutional choice. Benefits of institutional choice are more or less a function of crops grown. For plots with given crop in a locality, one can assume that there will not be much difference in benefits from irrigation if the technology and the use of other inputs in crop production are by and large similar. Under this condition, the net gain maximisation behaviour of the farmers could be reduced to a strategy of minimising the cost of acquiring irrigation.

A crude estimation of average cost associated with various institutions suggests that considerable variations exist across institutions. This apparent difference in costs highlights the need for a broader conceptualisation of costs, which also takes into account transaction costs, associated with these institutions. In irrigation, transaction costs are the costs of identifying profitable opportunities for acquiring water (information costs), costs of negotiating and bargaining or administratively deciding on the water transfer, the institutional costs of actually organising and contracting (formal and informal) and monitoring the transfers, and the cost of monitoring, enforcing and mitigating or eliminating possible third party effects and externalities.

With this wider conceptualistation, cost components of various institutions in the canal command were examined, which revealed the existence of substantial transaction costs across institutions. Thus, considering the existence of positive transaction costs associated with various institutional alternatives, economising takes place not with reference to production costs but with total costs, which is the sum of production and transaction costs.

Considering the importance of locational variables, the cost minimisation strategy of the farmers are analysed in terms of locational specificity of the plots. The sets of coordinates of locational factors leading to the selection of particular institutions were captured, which revealed concentration of institutions in specific locational trajectories. Locational attributes could affect cost in two ways. Firstly, these attributes could influence the cost associated with the setting up of the physical infrastructure of various institutions (production costs). Secondly, location would also have direct implications on the cost associated with organising, enforcing and monitoring various institutions (transaction costs).

The cost of physical infrastructure under public provision for all plots spread across the command area, irrespective of location, is marginal as these are set up by the State. Hence the only cost that is important in the context of institutional choice with respect to public provision is transaction cost. The fact that a farmer who is disadvantageously located does not opt for public provision thus indicates that the transaction costs associated with public provision in distant locations are higher. These higher transaction costs is found to result in the choice of alternative institutions in disadvantaged plots.

The production cost associated with collective action institutions, does not vary much across plots situated at different locations, as water is transferred mainly through the water-courses of the canal network. Here also, the only cost that could differ across farmers is the transaction costs. Unlike public provision and collective action in canal, alternative institutions incur substantial production costs. However, production cost associated with these institutions do not vary considerably across plots and hence, it is reasonable to assume that the selection of varied institutions in different plots are due to differences in transaction costs arising out of locational constraints. The empirical cases confirm that the net pay-off maximisation of the individual farmer is that of a cost minimisation problem, constrained by locational variables where transaction cost assumes a central concern.

Individual choices of institutions may not necessarily be efficient from the social point of view, which could lead to a gap between private and social gains. This is due to the unique features of irrigation such as : public good, externality and economies of scale. In the context of social efficiency due to the public good nature of irrigation, it was found that in institutions such as farmer initiated collective action, there are possibilities of non-cooperation and free riding by some of the beneficiaries. Anticipating this possibility, those who cooperate would contribute towards the institution only upto the point, where marginal costs equals the sum of their marginal willingness to pay (and not that of all users), which could lead to social inefficiency.

Irrigation due to its natural monopoly character, have potential for economies of scale. In the case of CRDS, if the gross additional cost incurred by farmers on different institutions would have been collected by an agency or government and channelised for the maintenance of the system, and if it results in increased supply and certainty, this implies that the *status quo,* characterised by a poorly functioning centralised system and a number of micro-institutions of water distribution is socially inefficient. Cost advantages are found to exist if a large number of farmers in a locality use the same institution of water distribution. The inability of these farmers to come together is found to lead to evolution of multiple institutions and thus increases the cost as a whole, leading to social inefficiency.

Externalities in irrigation arise when the institutional choice of certain irrigators lead to reduced availability of water to other farmers, who are also entitled to use the resource. Farmer's right to (quantum of) irrigation water are not well defined and inadequately enforced. This, coupled with the absence of volumetric pricing, lead to a situation where each farmer can use as much water as possible affecting the supply available to others. Appropriation of canal water (directly or indirectly) through various institutions located in the head reaches or other advantageously located positions, is found resulting in decreased water flow to tail-ends and, thus, affects the functioning of institutions at the tail-ends.

Equity considerations are important in the context of CRDS, as all the farmers are legally entitled to get irrigation water. Equity issues arise when the appropriation of water by the farmer through an institution reduces the availability to other potential users who have equal right to canal water or when benefits that accrued to an institutional change is not spread in a fair and just manner. The extraction of more water by strategically located farmers through alternative institutions, results in reduced supply of water to farmers located at disadvantageous positions leading to inequity. Institutions such as water market are found to have strong negative equity aspects with the possibility of rent seeking and skewed appropriation of gains.

Emerging Concerns

The existence of multiple institutions in irrigation raises questions on the efficiency and equity of public provision as well as that of the alternative institutions. The foremost issue in this regard is the growing inefficiency of public provision. This assumes increased concern in the context that the alternative institutions are mostly devised for transacting canal water. Thus, any inefficiency in canal irrigation is bound to have direct implications on the efficient functioning of alternative institutions.

Alternative institutions are resultant of the attempts to reduce costs by individual farmers. However, in the absence of unspecified boundaries and entitlements, existence of multiple institutions in the same canal command is bound to raise ambiguities and conflicts, leading to increases in costs. In this context, well-defined property rights and entitlements can bring in increased efficiencies in terms of reduced costs. This could also address many of the distributional issues in irrigation.

The distribution of irrigation water in the canal command through multiple institutions lead to higher costs, as there is no convergence of efforts between irrigators and Government. A rethinking and redefining the role of the State in distribution of water could alter the efficiency in provision to a greater extent. A major issue in this regard is the definition of the role and demarcation of boundaries of State involvement in distribution.

An important way to improve efficiency in water use is to increase the scarce value of water. This requires rethinking on the State policies on pricing irrigation as well as subsidising well, pump-sets and electricity. Irrigation is an input in agricultural production and hence has to be treated as an economic good, the price of which should be fixed on the basis of cost considerations. An equally important consideration in this regard is the realistic assessment of *ayacut* and the exemption of non-beneficiaries from the cess.

The efficiency of alternative irrigation institutions should also be looked into from the point of conjunctive water use, in the context of the increased relevance of promoting conjunctive water use, especially in the command area of canals. In the canal command of CRDS, alternative institutions are resorted to both for supplementary (conjunctive) and substitute source of irrigation. Well is found as the most important supplementry source used by most of the farmers, followed by water markets. Other institutions do not provide much scope for conjunctive use of water. Apparent difference exixts among different locations in terms of concentration of these institutions. Predominance of these institutions are observed in the middle reaches of the canal network as water availability is somewhat ensured (through seepage). In locationally disadvantaged plots, though private investments and water markets are seen, the uncertainty involved in getting seepage water is found affecting their spread. In this context, to enhance the overall efficiency, efforts could also be directed towards encouraging efficient and conjunctive use of water in the upper reaches through appropriate pricing of irrigation.

Policy Implications

The existence of multiple institutions in the provision of irrigation in the canal command obviously raises several concerns. The foremost of these is the failure of public provision and its sustainability as an institutional option in the distribution of irrigation. This also calls for a review of the assumption that Government is the right agency in the distribution of irrigation, till the reach of individual plots. As the cost of delivering water differs across users, depending on the location of the plots, providing irrigation at equal prices to all users needs to be relooked. The rationale of subsidising canal water-based on the assumption that farmers may not be able to pay for the service, is also contestable.

In this context, the prime requirement is to define the role and domain of state in canal irrigation management. As the sources of irrigation water are limited, its control could not be left entirely to the market or user groups due to chances of over use or excessive pricing of the resource. This attribute of irrigation as a natural monopoly warrants state's involvement. Further, large scale projects would also entail substantial economies of scale justifying large investments in irrigation. Moreover, large scale surface irrigation projects have substantial externalities and hence market forces and user groups are unlikely to take up this task. In this context, State's involvement in the construction and design of projects is desirable to promote investment in irrigation. Also, centralised control is strategically significant due to policy imperatives of the state, in the context of overall development of the economy.

Nevertheless, as public provision has proved to be a failure in the equitable distribution of water, it is appropriate to limit the state involvement to the construction and design of the system. Efficiency in the distribution of water can be ensured by tapping the possibilities of market/private and user group efforts, which can more effectively translate the irrigation requirements of the locality while deciding the distribution pattern. Possibilities of effecting judicious distribution through user groups are, however, limited as the costs of organising users and negotiating with them is high and it is most unlikely to sustain, without the support of state through interventions and subsidies. The advocacy for community based management of irrigation is often based on a neglect of opportunity cost of time and effort of members required to carry out and sustain such collective action. This has more implications in context of States, like Kerala, where majority of the farmers are part-timers and are involved in other economic activities. Under such cases, individual arrangements (private investment in wells) or market solutions (hiring of private agents or use of the services of a seller) may be appealing.

As water flows in the canal forms the basis of all the institutions, the crucial need for increasing efficiency lies in the proper operation and maintenance of the canal. This points towards the need for instituting an efficient and viable mechanism to recover at least, the recurring cost of canals. Irrigation needs to be treated as an input in agricultural production, which has opportunity cost and has to be priced with cost considerations that encourage more efficient resource use.

Alternative institutions, which are accepted by most of the farmers, are more likely to be productive and efficient than one that the state has designed, but which is considered unviable by the users. Thus, the only viable and cost effective solution to the problem of making irrigation benefits available to all the users at sustainable rates lies at strengthening the functioning of the alternative institutions and crafting mechanisms to regularise and legitimise these alternatives to ensure efficiency, equity and sustainability in irrigation management.

In the absence of unspecified boundaries, entitlements and property rights, market solutions are bound to raise ambiguities and conflicts, leading to an increase in transaction costs. In this context, state intervention is required to define and regulate well-defined, quantifiable and transferable property rights. The state should provide appropriate legal and institutional support in identifying, establishing and negotiating water rights and in monitoring and regulating externalities.

APPENDIX 1

Particulars of LBMC of CRDS

No	Name of Canal	I Crop (in ha.)	II Crop (in ha.)	III Crop (in ha.)	Total (in ha.)
1.	LBMC. km 0/0 to 11/200 km	65.09		81.34	146.43
	LBMC 11/2-18/4	128.20	152.00	148.13	428.33
	K.V Main	112.01	66.84	121.52	300.37
	Parakkadavu Main	291.64	176.08	209.45	677.17
	Kalady Main	115.00	155.49	297.01	567.50
2.	Adichily Branch	4234	35.65	68.29	146.28
3.	Bhoothamkutty Branch	24.30	19.62	34.40	78.32
4.	Meloor Branch	50.35	70.60	68.72	189.67
5.	Poolani Branch	52.64	21.04	61.21	134.89
6.	Thanguchira Branch	92.53	38.40	37.97	168.80
7.	Kizhakkummuri Branch	46.10	36.20	58.83	141.13
8.	Chirangara Branch	27.00	26.10	41.10	94.20
9.	Meloor South Branch and Distributary	156.81	81.32	227.93	466.06
10.	Koratty Branch	34.72	30.22	56.62	121.56
11.	Konoor Branch	2.10	6.40	9.33	23.83
12.	M.V Branch Canal	52.50	38.16	74.46	165.42
13.	K.V Branch	60.47	45.49	82.09	188.05
14.	Edakunny Branch	60.46	46.04	69.70	176.20
15.	Marangadam Branch	35.60	20.40	44.32	100.32
16.	Karukutty-Karayamparambu Branch	121.69	73.41	168.09	363.19
17.	Mambra Branch	60.16	48.10	80.66	188.92
18.	Peechanikkad Branch	56.10	35.63	68.76	160.40
19.	Parakkadavu Branch	36.47	31.28	40.38	108.13
20.	Attara Branch	42.30	39.65	69.68	151.63
21.	Azhakam Branch	41.43	32.33	61.27	135.03
22.	Karukuttykara Branch	98.00	41.36	58.75	198.11
23.	Manjapra Branch and Distributary	251.31	150.24	286.98	688.53
24.	Anapara Branch	18.40	12.69	24.55	55.64
25.	Naduvattam Branch	25.67	30.66	36.75	93.08
26.	Bhoothamkutty South Branch	22.04	26.10	38.09	96.22
27.	Kalady Branch	50.26	31.63	69.27	151.16
28.	Komarapadam Branch	91.00	41.06	58.01	190.07
29.	Vengoor Branch	90.20	45.68	142.82	278.70
30.	Thottakam Branch	30.41	22.47	46.01	98.89
31.	Chengamanad Branch	20.33	15.40	29.99	65.72
	Total	2521.63	1743.74	3072.77	7338.14

Source : *Preliminary Project Report on CRDS in Thrissur and Ernakulam Districts in Kerala*, Report submitted to National Water Management Programme, Project Wing-Irrigation Department, Government of Kerala, 1994.

APPENDIX 2

Hydraulic Particulars of LBMC of CRDS

No.	Name of Canal	Branches at	Length	Bed Slope	Bed Width	Side Slope	F.S.D.	Velocity m^3/sec	Discharge in m^3/sec
1.	LBMC. km 0/0 to 11/200 km		11.200		6.10		2.00	0.97	12.75
2.	Adichily Branch	4/100 of LBMC	7.700		1.37		0.52	0.24	0.24
3.	Bhoothamkutty Branch	9/100 of LBMC	6.400		1.37		0.61	0.34	0.18
4.	Meloor Branch	10/200 of LBMC	6.437		1.83		0.61	0.31	0.42
5.	Poolani Branch	6/500 Meloor Branch	3.218		1.37	½ : 1	0.53	0.34	0.22
6.	Thanguchira Branch	11/200 of LBMC	4.525		1.07	½ : 1	0.53	0.17	0.14
7.	LBMC. km11/200 to 18/400		7.200	11/2-14	4.10				
				14-19/4	3.35	½ : 1	1.22	0.70	3.22
				17/460 to 18/400	3.05				
8.	K.V Main	14/000 of LBMC	6.600	0/0 to 2/184	2.75				
				2/184 to 6/600	2.13	½ : 1	0.90	0.47	1.53
9.	Kizhakkummuri Branch	14/200 of LBMC	2.816		0.90	½ : 1	0.61	0.28	0.18
10.	Chirangara Branch	20/400 of LBMC	1.006		0.90	½ : 1	0.46	0.18	0.110

(Contd. Appendix 2)

(APPENDIX 2 contd.)

No.	Name of Canal	Branches at	Length	Bed Slope	Bed Width	Side Slope	F.S.D	Velocity m³/sec	Discharge in m³/sec
11.	Meloor South Branch	2/184 KV Main	4.968	0/0 to 3/24	1.31	½ : 1	0.60	0.20	0.45
				3/24 to 4/568					
12.	Meloor South Dist.	3/324 of Meloor	1.609		0.90	½ : 1	0.61	0.20	0.18
13.	Koratty Branch	4/0 of KV Main	1.207		0.90	½ : 1	0.46	0.18	0.11
14.	Konoor Branch	4/400 of KV Main	1.550		0.75	½ : 1	0.30	0.10	0.30
15.	K.V Branch	6/630 of KV Main	4.627		1.68				
					1.20	½ : 1	0.61	0.52	0.41
16.	M.V Branch Canal	5/630 of K V Main	4.425	0/0 to 1/410	1.98				
				1/410 to 3/01	1.85				
				3/01 to 4/255	0.90	½ : 1	0.61	0.31	0.49
17.	Edakunny Branch	13/600 of LBMC	3.600	0/00 to 1/320	1.37				
				1/320 to 3/600	0.90	½ : 1	0.46	0.49	0.38
18.	Marangadam Branch	17/460 of LBMC	1.400		0.90	½ : 1	0.46	0.37	0.23
19.	Parakkadavu Main—I reach—km18/4-25/2	18/400	6.30		3.30	½ : 1	1.05	0.70	1.48
20.	Parakkadavu Main—II reach—km 25/2-23/4	25/200	3.20		2.50	½ : 1	0.75	0.45	0.62
21.	Parakkadavu Main—III reach—km 28/4—33/2	28/4	4.80		1.20	½ : 1	0.45		
22.	Karukutty-Karayamparambu Branch	18/540 of Para-kkadavu Main	6.200	0/0 to 3/300	2.00	1½ : 1	0.61	0.33	0.53
				3/300 to 6/2					
23.	Mambra Branch	25/330 of Para-kkadavu Main	2.970		1.22	1½ : 1	0.61	0.55	0.56

24.	Peechanikkad Branch	27/670 of Parakadavu Main	3.219		1.20	1½ : 1	0.46	0.45	0.35
25.	Parakadavu Branch	28/220 of Parakadavu Main	3.319		1.20	1½ : 1	0.38	0.48	0.35
26.	Attara Branch	3/600 of Kalady Main	1.800		0.90	1½ : 1	0.56	0.43	0.35
27.	Azhakam Branch	6/200 of Kalady Main	1.400		0.90	1½ : 1	0.38	0.41	0.20
28.	Karukuttykara Branch	5/600 of Kalady Main	2.600		0.90	1½ : 1	0.53	0.54	0.41
29.	Kalady Main—0/0 to 12/4		12.400		4.00				
					3.30				
					2.75	1½ : 1	1.07	0.71	3.84
30.	Manjapara Branch	4/200 of Kalady Main	8.600		2.50				
					1.85				
					1.20	1½ : 1	0.90	0.72	0.90
31.	Anapara Branch	2/600 of Manjapra Branch	1.300		0.90	1½ : 1	0.46	0.22	0.14
32.	Naduvattam Branch	6/800 of Manjapra Branch	2.339		0.90	1½ : 1	0.46	0.39	0.19
33.	Bhoothamkutty South Branch	3/600 of Kalady Main	1.700	0/0 to 3/6	0.90	1½ : 1	0.46	0.30	0.19
34.	Kalady Branch	12/400 of Kalady Main	5.200	3/6 to 5/2	2.80				
					0.90	1½ : 1	0.46	0.29	0.39
35.	Manjapra Distributary	3/200 of Manjapara Branch	2.167		0.90	1½ : 1	0.46	0.30	0.19

(*Contd.*)

(APPENDIX 2 contd.)

No.	Name of Canal	Branches at	Length	Bed Slope	Bed Width	Side Slope	F.S.D	Velocity m³/sec	Discharge in m³/sec
36.	Komarapadam Branch	5/00 of Kalady Main	4.00	0/0 to 2/0	0.90	1½ : 1	0.53	0.54	0.41
37.	Vengoor Branch	9/400 of Kalady Main	6.00	2/0 to 2/2	1.85				
					1.20	1½ : 1	0.61	0.31	0.47
38.	Thottakam Branch	12/400 of Kalady Main	2.600	2/2 to 6/0	0.90	1½ : 1	0.46	0.35	0.22
39.	Chengamanad Branch	3/600 of Kalady Main	1.600		1.83	1½ : 1	0.46	0.20	0.21

Source: *Preliminary Project Report on CRDS in Thrissur and Ernakulam Districts in Kerala*, submitted to National Water Management Programme : Government of Kerala, Irrigation Department-Project Wing, 1994.

APPENDIX 3

Questionnaire

Institutional Choice in Irrigation : A Study of Distribution in a Command Area in Kerala

Sample No. :
Date of Interview :

A. General Schedule
A.I. Household Identification :

(1) Name of the household head :

(2) Name of the respondent :

(3) Relation to the household head :

(4) House No./Ward No./Panchayat/Village :

(5) Religion/Caste/Community :

(6) Occupational status of the head :
Main Subsidiary
(1)
(2)
(3)

(7) Approximate Annual Income :
Main Subsidiary
(1)
(2)
(3)

Total Annual Income :

(8) Socio-economic Characteristics of the Household Members :

Sl. No.	Name	Age	Sex	Relationship to the Head	Education	Employment	Occupation		Approximate Annual Income	
							Main	Sub	Main	Sub

Sex	Employment	Occupation	Relation to Head	Education
1. Male 2. Female	1. Self employed 2. Regular employment (salaried) 3. Casual employment	1. Cultivator 2. Agri. Labour 3. Cattle keeping 4. Non-Agri. Labour 5. Trade, commerce, transport service 6. Student 7. Domestic work 8. Govt. Employment 9. Others (Specify)	1. Father 2. Mother 3. Wife 4. Son 5. Daughter 6. In-law 7. Grand child 8. Others	1. Illiterate 2. Upto SSLC 3. SSLC—Graduation 4. Graduation and above

A.II Details of Land

(1) General

Plot No.	Year of Possession	Category	Distance from Residence (in metres)	Type	Soil	Total Area	Cultivated Area	Fallow
1.								
2.								
3.								
4.								
5.								

Type	*Category*
1. Dry	1. Owned and operated
2. Wet	2. Leased-in
	3. Leased-out

(2) Season-wise Details of Land Use

Plot No.		Cultivated Area			Fallow	Source of Irrigation	Cropping Type	Crops Grown	
		Irrigated	Unirrigated	Total				Main	Sub
1.	S1								
	S2								
	S3								
2.	S1								
	S2								
	S3								
3.	S1								
	S2								
	S3								
4.	S1								
	S2								
	S3								
5.	S1								
	S2								
	S3								

S1 : June—September *Cropping Type*
S2 : October—January 1 Mono cropping
S3 : February—May 2 Mixed cropping

(3) Were you following the same cropping pattern before? Yes/No

(4) If No,
(a) When did you change the cropping pattern?

(b) Details of Previous Land Use.

Plot No.		Cultivated Area			Fallow	Source of Irrigation	Cropping Type	Crops Grown	
		Irrigated	Unirrigated	Total				Main	Sub
1.	S1								
	S2								
	S3								
2.	S1								
	S2								
	S3								
3.	S1								
	S2								
	S3								
4.	S1								
	S2								
	S3								
5.	S1								
	S2								
	S3								

S1 : June—Sept.; S2 : Oct.—January; S3 : February—May
Cropping Type : 1. Mono Cropping, 2. Mixed Cropping

(c) Reasons for the shift in cropping pattern.
   (i) Increased cost of cultivation
   (ii) Low profitability
   (iii) Labour shortage
   (iv) Water scarcity
   (v) Others (Specify)

(5) Details on crop pattern history
   (a) Approximate year of shifts
   (b) Crop pattern
   (c) Reasons for shift

A. III. Crop-wise Details

(1) Distribution of Total Area under Different Crops.

Crops Grown	*Area Under*															*Total Area*		
	Plot 1			*Plot 2*			*Plot 3*			*Plot 4*			*Plot 5*			*S1*	*S2*	*S3*
	S1	*S2*	*S3*	*S1*	*S2*	*S3*	*S1*	*S2*	*S3*	*S1*	*S2*	*S3*	*S1*	*S2*	*S3*			
Paddy																		
Coconut																		
Arecanut																		
Rubber																		
Tapioca																		
Plantain																		
Nutmeg																		
Vegetables																		
Others, specify																		
Total area																		

S1 : June—Sept.; S2 : Oct.—Jan.; S3 : February—May

(2) Details of Crop Rotation (if practiced)

Plot No.	*Details on Rotation of Crops*
1.	
2.	
3.	
4.	
5.	

(3) Details of Irrigation

Plot No	*Main Crops*	*Number of Irrigation*		*Frequency of Irrigation*		*Hours of Irrigation*	
		Actual	*Required*	*Actual*	*Required*	*Actual*	*Required*
1.							
2.							
3.							
4.							
5.							

B. Plot-wise Details on Irrigation Institutions
B.I. Canal Irrigation

(1) Year of commencement of canal irrigation :
(2) Name of the branch :
(3) Location in the branch :
(4) Approximate distance from the nearest outlet :
(5) Approximate distance from the nearby branch canal :
(6) Approximate elevation of the plot from the water in the canal :
(7) Actual timing and duration of supply :
(8) Crops irrigated using canal water :
(9) Extent of water availability : Adequate/Inadequate/No supply
(10) If Inadequate/No supply :
  (i) From when :

  (ii) Reasons :
    (a) Excessive use of water by head/middle end farmers
    (b) Insufficient water release into the canal
    (c) Wastage of water due to bad maintenance of the canal
    (d) Farmers at the head end are economically and politically powerful and *over appropriate irrigation
    (e) Unauthorised cultivation by head-end farmers
    (f) Extension of canal
    (g) Other reasons, specify
  (iii) Measures taken to face water scarcity :
    (a) Action for the improvement of the system (specify)
    (b) Alternative arrangements (specify)
    (c) Crop pattern change (specify)
    (d) Other land use change
    (e) Bribing
    (f) Influencing

(11) If there were any steps for the improvement of the system, specify whether it was Government induced or farmer motivated?

(12) Are there any canals passing near your field : Yes/No

(13) If yes, how it helps or affects your cultivation?

(i) Able to irrigate land without formal release of water

(ii) Land gets flooded because of seepage of water

(iii) Causes soil erosion and land degradation

(iv) Leads to ground-water recharge and helps to irrigate during non-availability of water in the canals

(v) Loss of water for cultivation

(14) Are any of your lands/part of your lands are waterlogged or have become saline due to canal? Yes/No

(15) Are you satisfied with the overall performance of the canal irrigation project? Yes/No

(16) Remarks on the performance of canal :

No.	Items	Highly Satis-factory	Satis-factory	Not Satis-factory	Remarks
1.	Time of opening the canal				
2.	Time of closing the canal				
3.	Water flow				
4.	Maintenance of canal				
5.	Maintenance of field channel				
6.	Irrigation charges				
7.	Attending complaints				
8.	Behaviour of staff				
9.	Settling of disputes				

(17) Have you ever brought the problem of water scarcity to the notice of the irrigation authorities?

(18) If yes, what was the response?

(19) Have your crops faced any damage due to canal water distribution? Yes/No

(20) If yes, what are the problems? Give details

The Problem	Crop	Season	At what stage of crop growth	Extent of damage to the crop

(21) The nature of conflicts in canal irrigation if any?

(22) `How do you solve these conflicts?

(23) Total cost involved in getting canal irrigation.
 (i) Government levy,
 (ii) Money spend other than government levy,
 (iii) Time and effort spend in getting the spout open, and in solving disputes.

(24) Suggestions for improving the performance of canal irrigation.

(25) Does your plot come under the operation area of CADA? Yes/No.

(26) Are you satisfied with the functioning of CADA? Yes/No

(27) If yes, what are the benefits of CADA?

(28) What are the crops irrigated with the help of CADA?

(29) Suggestions to improve the working of CADA.

B.II. Government Initiated Collective Action in Canal Irrigation :

(1) Are you a member of CADA Committee? Yes/No.

(2) If yes, what is the nature of collective action?

(3) When did you join this institution?

(4) What are the reasons for going for CADA Committee membership?

(5) What are the crops irrigated under this institution?

(6) Number of farmers in the committee :

Large | Active
Medium | Passive
Small

(7) Details on the organisation of the committee :

(8) The cost associated with the membership :
(1) In terms of money :
(2) In terms of mandays

(9) Is there any change in land use or cropping pattern after becoming a member of this institution?

(10) Are you satisfied with the functioning of CADA Committee? Yes/No

(11) If yes, reasons

(12) If no, reasons

(13) Farmer response on the performance CADA Committee .

No	Items	Highly Satisfactory	Satisfactory	Not Satisfactory	Remarks
1.	Maintenance of canal				
2.	Maintenance of field channel				
3.	Attending complaints				
4.	Settling of disputes				
5.	Improvement in supply				
6.	Organisation of the committee				

(14) Suggestions to improve the performance of CADA Committee

B. III. Farmer Initiated Collective Actions in Canal Irrigation :

(1) Are you a member in any collective action ventures? Yes/No.

(2) If yes, what is the nature of collective action?
(i) Collective bribing,
(ii) Collective pressurising and bargaining,
(iii) Collective action in operation and maintenance,
(iv) Collective action in solving disputes,
(v) Other, specify.

(3) When did you join this institution?

(4) What are the reasons for going for collective action?

(5) What are the crops irrigated under this institution?

(6) Number of farmers in group-action :

Large farmers :	Active Members :
Medium farmers :	Passive Members :
Small farmers :	
Total :	

(7) The cost associated with the membership :
(1) In terms of money
(2) In terms of mandays
(3) In terms of efforts

(8) Is there any change in land use or cropping pattern after becoming a member of this institution?

(9) Are you satisfied with the functioning of the institution? Yes/No.

(10) If yes, reasons.

(11) If no, reasons.

(12) Farmer response on the performance of collective action.

No	Items	Highly Satisfactory	Satisfactory	Not Satisfactory	Remarks
1.	Time of opening the canal				
2.	Time of closing the canal				
3.	Water flow				
4.	Maintenance of canal				
5.	Maintenance of field channel				
6.	Attending complaints				
7.	Settling of disputes				

(13) Suggestions to improve the performance.

B. IV. Other Institutions

(1) Further to canal irrigation, do you make use of some other sources for irrigating your crop? Yes/No.

(2) If yes, give details.

<table>
<tr><th rowspan="3">No.</th><th rowspan="3">Source of Irrigation</th><th colspan="3">Nature of Source</th><th rowspan="3">Own</th><th colspan="2">Crops Grown</th><th rowspan="3">Area (acres)</th></tr>
<tr><th rowspan="2">Own</th><th colspan="2">Otherwise</th><th colspan="2">Otherwise</th></tr>
<tr><th>Water Market</th><th>Community Irrigation</th><th>Water Market</th><th>Community Irrigation</th></tr>
<tr><td>1.</td><td>Irrigation by wells</td><td></td><td></td><td></td><td></td><td></td><td></td><td></td></tr>
<tr><td>2.</td><td>Irrigation by others</td><td></td><td></td><td></td><td></td><td></td><td></td><td></td></tr>
<tr><td>3.</td><td>Total</td><td></td><td></td><td></td><td></td><td></td><td></td><td></td></tr>
</table>

(3) Why are you resorting to such sources? Give reasons.
(i) Water scarcity in canal,
(ii) Risk/uncertainty minimisation,
(iii) Cost minimisation,
(iv) Locational advantage or disadvantage,
(v) Any other (specify).

(4) Crops irrigated under the present institution :

(5) Source of water under the present institution :

(6) Is there any change in land use or cropping pattern after becoming a member of this institution?

(7) Do you have any plans to go back to canal irrigation? Yes/No.

(8) Reasons.

B. IVa. Details on irrigation by wells :

(1) No of wells,

(2) Constructed/already existing,

(3) Source of water in the well :
(i) Exclusively ground water,
(ii) Seepage water.

(4) Are you having more water in the well than earlier? Yes/No.

(5) If yes, from when?

(6) Subsidy in digging the well :

(7) Year of digging the well :

(8) Type of pump-set with HP :

(9) Subsidy in pump set :

(10) Cost associated with irrigation from well :

No. Electrified/ Kerosene pump set (HP)	*Fixed Cost*		*Recurring Cost*		*Total Cost*	
	Monetary	*Labour*	*Monetary*	*Labour*	*Monetary*	*Labour*
1.						
2.						
3.						

(11) Are you satisfied with the functioning of the institution? Yes/No.

(12) Reasons.

B.IVb. Irrigation by Water Markets :

(1) What is the source of water in the well?
(i) Exclusively ground water,
(ii) Seepage water.

(2) What are the terms and conditions of supply?
(i) Fixed charges,
(ii) Charge per acre/hour,
(iii) The supply conditions.

(3) Type of pump-set with HP.

(4) If the present arrangement is through negotiation how much time you have spent on it?

(5) Are you satisfied with the functioning of the institution? Yes/No.

(6) Reasons.

B.IVc. Irrigation by other Sources : Ponds, River etc. : Individual :

(1) What is the source of water?
(1) Exclusively ground water,
(2) Seepage water,
(3) River water.

(2) Cost associated :

No.	Electrified/ Kerosene pump set (HP)	Fixed Cost		Recurring Cost		Total Cost	
		Monetary	Labour	Monetary	Labour	Monetary	Labour
1.							
2.							
3.							

(3) Are you satisfied with the functioning of the institution? Yes/No.

(4) Reasons.

B. IVd. Irrigation by Other Sources : Ponds, River etc. : Water Markets :

(1) What is the source of water?
(i) Exclusively ground water,
(ii) Seepage water,
(iii) River water.

(2) What are the terms and conditions of supply?
(i) Fixed charges,
(ii) Price per acre/hour,
(iii) The supply conditions.

(3) Type of pump-set with HP.

(4) If the present arrangement is through negotiation how much time you have spent on it?

(5) Are you satisfied with the functioning of the institution? Yes/No.

(6) Reasons.

B. IVe. Irrigation by Other Sources : Well, Ponds, River etc. : Community Irrigation :

(1) What is the source of water?
(i) Exclusively ground water,
(ii) Seepage water,
(iii) River water,
(iv) Canal water.

(2) Number of members

Large farmers :	Active Members :
Medium farmers :	Passive Members :
Small farmers :	
Total :	

(3) Type of pump set with HP,

(4) Extent of subsidy involved,

(5) Subsidising Agency,

(6) Total initial cost,

(7) Total recurring cost,

(8) The cost associated with the membership.
(i) in terms of money,
(ii) in terms of mandays,
(iii) in terms of efforts.

(9) Are you satisfied with the functioning of the institution? Yes/No.

(10) Reasons.

B.V. Details on Previous institution :

(1) The institutions in which the farmer was a member earlier :

Institution	*Duration*	*Crops Grown*		*Cost*				*Reasons to quit the institution*
		Main	*Sub*	*Fixed Cost*		*Recurring Cost*		
				Monetary	*Labour*	*Monetary*	*Labour*	

B.VI. Others :

(1) Whether a shift to some other institution will improve the situation; if so?
(i) The institution that you think superior :
(a) Economically,
(b) Risk-wise.

(2) The reasons that hinder from joining that institution :

C. Cost of cultivation details :

(1) Production and cost of cultivation details : Paddy

Plot No.	Land type	Area	No of crops grown in an year	Cost of Cultivation excluding irrigation					Total Yield		Marketed Output	
				Land deve-lop-ment	Labour	Seed	Manure Fertili-ser and Pesti-cide	Harvesi-ing and Process-ing	Paddy	Hay	Paddy	Hay
1.												
2.												
3.												
4.												
5.												

Type : 1. Wet; 2. Dry

(2) Production and cost of cultivation details : Coconut

Plot No.	Land Type	Area/ No. of Plants	Mixed or Mono	If Mixed other Crops	Cost of Cultivation			Total Yield	Marketed Output
					Labour	Manure, Fertiliser and Pesticide	Harvesting and Processing		
1.									
2.									
3.									
4.									
5.									

Type : 1. Wet; 2. Dry

(3) Production and cost of cultivation details : Areca

Plot No.	Land Type	Area/ No. of Plants	Mixed or Mono	If Mixed other Crops	Cost of Cultivation			Total Yield	Marketed Output
					Labour	Manure, Fertiliser and Pesticide	Harvesting and Processing		
1.									
2.									
3.									
4.									
5.									

Type : 1. Wet; 2. Dry.

(4) Production and cost of cultivation details : Nutmeg

Plot No.	Land Type	Area/ No. of Plants	Mixed or Mono	If Mixed other Crops	Cost of Cultivation			Total Yield	Marketed Output
					Labour	Manure, Fertiliser and Pesticide	Harvesting and Processing		
1.									
2.									
3.									
4.									
5.									

Type : 1. Wet; 2. Dry.

(5) Production and cost of cultivation details : Rubber

Plot No.	Land Type	Area/ No. of Plants	Mixed or Mono	If Mixed other Crops	Cost of Cultivation			Total Yield	Marketed Output
					Labour	Manure, Fertiliser and Pesticide	Harvesting and Processing		
1.									
2.									
3.									
4.									
5.									

Type : 1. Wet; 2. Dry.

(6) Production and cultivation details of other crops : Specify

Plot No.	Land Type	Area/ No. of Plants	Mixed or Mono	If Mixed other Crops	Cost of Cultivation			Total Yield	Marketed Output
					Labour	Manure, Fertiliser and Pesticide	Harvesting and Processing		
1.									
2.									
3.									
4.									
5.									

Type : 1. Wet; 2. Dry.

D. Any Other Remarks :

Bibliography

Abernethy, Charles, L. (ed.), 1996. The Institutional Framework for Irrigation, Workshop Proceedings, International Irrigation Management Institute, Colombo.

Abhayaratna, M.D.C.; Vermillion, S.; Johnson and Perry, C. (ed.), 1994. Farmer Management in Ground Water Management in Asia, Selected Papers from a South Indian Regional Workshop, International Irrigation Management Institute, Colombo.

Akerlof, George, A., 1970. The Market for Lemons : Qualitative Uncertainty and Market Mechanism, *Quarterly Journal of Economics*, Vol. 84. No. 3.

Alchian, A. and Demsetz, H., 1972. Production, Information Costs and Economic Organisation, *American Economic Review*, Vol. 62, No. 5.

Alchian, A. and Demsetz, H., 1973. The Property Rights Paradigm, *Journal of Economic History*, Vol. 33, No. 5.

Arrow, 1969. The Organisation of Economic Activity : Issues Pertinent to the Choice of Market *versus* Non-market Allocation in the Analysis and Evaluation of Public Expenditure : The PPB System—Vol. 1., U.S Joint Economic Committee, 91st Congress, I Session, Washington, D.C.

Arthur, W. Brian, 1989. Competing Technologies, Increasing Returns, and Lock-in by Historical Events. *Economic Journal*, Vol. 99, No. 39.

Ballabh, Vishwa and Shunmugam, V., 1998 Distributed Governance in Basin Management—A Paradisaical Denoument, *The Social Engineer*, Vol. 7 No. 2.

Bandaragoda, D.J., 1998. Design and Practice of Water Allocation Rules : Lessons From Warabandi in Pakistan's Punjab, Research Report 17, International Water Management Institute, Colombo.

Bandaragoda, D.J., 1999. Institutional Change and Shared Management of Water Resources, in *Large Canal Systems : Results of an Action Research Programme in Pakistan*, Research Report 36, International Water Management Institute, Colombo.

Bandaragoda, D.J., 2000. A Framework for Institutional Analysis for Water Resource Management in a River Basin Context, Working Paper 5, International Water Management Institute, Colombo.

Bandaragoda, D.J. and Rehman, Saeed Ur., 1994. Institutional Perspectives of Land Reclamation Operation in Punjab, Working Paper, 28, International Irrigation Management Institute, Colombo.

Bandaragoda, D.J. and Rehman, Saeed Ur., 1995. Warabandi in Pakistan's Canal Irrigation Systems : Widening Gap between Theory and Practice, IIMI Country Paper, Pakistan No. 7, International Irrigation Management Institute, Colombo.

Bandyopadhyay, S. and von Eschen, D., 1988. Village Failure to Cooperate : Some Evidence from West Bengal, India in Attwood, D.W and Baviskar, B.S (ed.) *Who Shares? Cooperatives and Rural Development*, New Delhi : Oxford University Press.

Bardhan, Pranab, 1993. Analytics of the Institution of Informal Co-operatives in Rural Development. *World Development*, Vol. 21, No. 4.

Bardhan, Pranab, 2000. Irrigation and Co-operation : An Empirical Analysis of 48 Irrigation Communities in South India, *Economic Development and Cultural Change*, Vol. 48, No. 4.

Boyce, J., 1987. Agrarian Impasse in Bengal : Institutional Constraints to Technological Change, Oxford University Press, Oxford

Brewer, J.; Kolavalli, S.; Kalro, A.H.; Naik, G.; Ramnarayan, S.; Raju, KV. and Sakthivadivel, R., 1999a. Irrigation Management Transfer in India : Policies, Processes and Performance, Oxford and IBH Publishing Company, New Delhi.

Brewer, J.D., Shaktivadivel. R. and Raju K.V. 1999b. Water Distribution Rules and Water Distribution Performance : A Case Study in the Tambraparni Irrigation System. IWMI Research Report 12, International Water Management Institute, Colombo.

Bromley, Daniel, W. 1989. Economic Interests and Institutions : The Conceptual Foundations of Pubilc Policy, Basil Blackwell, New York.

Bromley, Daniel, W.; Taylor, D.C. and Parker, D.E., 1980. Water Reform and Economic Development : Institutional Aspects of Water Management in the Developing Countries, *Economic Development and Cultural Change,* Vol. 28, No. 2.

Casey, Margaret, 1991. Ground Water Development in Madura, Indonesia : Gender Issues in an Irrigation Project, Network Paper 1, ODI Irrigation Management Network, London.

Chackacherry, G., 1993. Farmer Associations and Utilisation of Water : A Case Study from Kerala State, India in *Proceedings of the International Conference on Environmentally Sound Water Resources Utilisation,* November 8-11, Bangkok.

Chambers, Robert, 1988. Managing Canal Irrigation : Practical Analysis from South Asia, Cambridge University Press, New York.

Chopra, K.; Kadekodi, G.K. and Murthy, M.N., 1989 : Peoples' Participation and Common Property Resources, *Economic and Political Weekly,* Vol. XXV, Nos. 51-52.

Chopra, K.; Kadekodi, G.K. and Murthy, M.N., 1990. Participatory Development : People and Common Property Resources, New Delhi : Sage Publications.

Clague, Christopher, 1997. Institutions and Economic Development : Growth and Governance in Less Developed and Post-Socialist Countries, The John Hopkins University Press, London.

Coase, R.H., 1960. The Problem of Social Cost, *Journal of Law Economics,* Vol. 3.

Coward Jr. and Uphoff , N. 1986. Operation and Maintenance in Asian Irrigation : Repressing Government and Farmer Responsibilities and Rights, *Irrigation and Drainage Systems,* Vol. 1.

Dasgupta, P.S and G.M. Heal, 1979. Economic Theory and Exhaustible Resources, Cambridge University Press, London.

David, Paul, 1985. CLIO and Economics of QWERTY. *American Economic Review*, Vol. 75, No. 2.

Demsetz, H., 1967. Toward a Theory of Property Rights, *American Economic Review*, Vol. 62.

Dhawan, B.D., 1985. Irrigation Impact on Farm Economy, *Economic and Political Weekly*, Vol. XX, No. 39.

Dhawan, B.D., 1988. Indian Irrigation : An Assessment, *Economic and Political Weekly*, Vol. XXIII, No. 19.

Dhawan, B.D., 1997. Large Scale Canal Irrigation : How Cost Effective?, *Economic and Political Weekly*, Vol. XXXII, No. 26.

Dhawan, B.D. and Datta H.S., 1992. Impact of Irrigation on Multiple Cropping, *Economic and Political Weekly*, Vol. XXVII, No. 13.

Domenich, A. and McFadden, D., 1975. Urban Travel Demand : A Behavioural Analysis, North-Holland American Elsevier, Amsterdam.

Duane, Paul, 1975. Policy Framework for Irrigation Water Charges. World Bank Staff Working Paper 48. World Bank, Washington D.C.

Eggertsson, Thrainn, 1990. Economic Behaviour and Institutions, Cambridge University Press, Cambridge.

Foldvary, Fred, 1995. Public Goods and Private Communities : The Market Provision of Social Services, Edward Elgar Publishing Limited, England.

Francis, S.K., 1990. Dynamics of Rural Labour Markets : An Analysis of Agricultural Labour Shortage in a Kerala Region. Unpublished M. Phil. Thesis, Centre for Development Studies, Trivandrum.

George, M.V. and Nair N.G., 1982. Irrigation and Agriculture Development in Kerala. In Pillai, P.P. (ed) *Agriculture Development in Kerala*, Agricole Publishing Company, New Delhi.

George, P.S., 1982. Agricultural Price Movements in Kerala in Agriculture Development. In Pillai, P.P. (ed) *Agriculture Development in Kerala*, Agricole Publishing Company, New Delhi.

Government of India, 1987. National Water Policy. Ministry of Water Resources, New Delhi.

Government of Kerala, 1994. Preliminary Project Report on Chalakkudy River Diversion Scheme in Thrissur and Ernakulam Districts in Kerala. Irrigation Department, Trivandrum.

Greene, William H., 1993. Econometric Analysis, MacMillian, London.

Hardin, Garett, 1968. The Tragedy of Commons. *Science*, Vol. 162.

Hubbard, Michael, 1997. The New Institutional Economics in Agricultural Development : Insights and Challenges, *Journal of Agricultural Economics*, Vol. 48, No. 2.

IIMI, 1997. Water Rights, Conflicts and Policy : Proceedings of a Workshop, International Irrigation Management Institute, Colombo.

Jacob, J., 1996. Institutional Interventions in Water Management : A Micro Level Study of Organisation of Irrigation in Kerala, Project Paper, Centre for Development Studies, Trivandrum.

Jairath, Jasveen, 1985. Technical and Institutional Factors in Utilisation of Irrigation : A Case Study of Public Canals in Punjab, *Economic and Political Weekly*, Vol. XX, No. 13.

Janakarajan S., 1994. Trading in Ground Water : A Source of Power and Accumulation. Working Paper 123, Madras Institute of Development Studies, Madras.

Jayaraman, T.K., 1981. Farmers' Organisations in Surface Irrigation Projects : Two Empirical Studies from Gujarat. *Economic and Political Weekly*, Vol. XVI, No. 39.

Jong, Ijsbrand H. de, 1989. Fair and Unfair : A Study into the Bathma System in Two Sri Lankan Village Irrigation Systems, Working Paper 15, International Irrigation Management Institute, Colombo.

Jose, S., 1991. Group Farming in Kerala : An Illustrative Study, Unpublished M.Phil Thesis, Centre for Development Studies, Trivandrum.

Joshi L.K and Hooja, R., 2000. Participatory Irrigation Management : Paradigm for 21st Century, Vols. 1 and 2, Rawat Publication, New Delhi.

Kannan, K.P. and Pushpangadhan, K., 1988. Agricultural Stagnation in Kerala : An Exploratory Analysis. *Economic and Political Weekly* , Vol. XXIII, No. 39.

Kannan, K.P. and Pushpangadhan, K., 1989. Agricultural Stagnation and Irrigation in Kerala. *Economic and Political Weekly*, Vol. XXIV, No. 19.

Kiser, Larry L. and Ostrom E., 1982. The Three Worlds of Action. A Meta-theoretical Synthesis of Institutional Approaches.in Ostrom E (ed) *Strategies of Political Inquiry*, Sage California.

Klitgaard, Robert, 1996. Comment on 'Incentives, Rules of the Game, and Development'. in Breno-Michael and Pleskovic, Bris (ed) *Annual World Bank Conference on Development Economics*. World Bank, Washington, D.C.

Kollavalli, Shashi, 1995. Conditions of Collective Action : An Application to Water User Associations, mimeo, Indian Institute of Management, Ahmedabad.

Kollavalli, Shashi, 1996. User Participation and Incentives to Perform in Indian Irrigation Systems, Centre for Management in Agriculture, Indian Institute of Management, Ahmedabad.

Kollavalli, Shashi and Shah, Nithin, 1989. Management of Public Tube Wells in U.P., Indian Institute of Management, Ahmedabad.

Kollavalli, Shashi and Kerr, John, 2002. Mainstreaming Participatory Watershed Development, *Economic and Political Weekly*, Vol. XXXVII, No. 3.

Lam, Fung W.; Lee, M. and Ostrom, E., 1997. The Institutional Analysis and Development Framework : Application to Irrigation Policy in Nepal, Policy Studies and Developing Nations, JAI Press Inc.

Libecap, 1978. Economic Variables and the Development of the Law : The Case of Western Mineral Rights, *Journal of Economic History*, Vol. XXXVIII, No. 2.

Long, Roger, B., 1983. Measuring the Contribution of Land and Water to Agriculture, *Water Resource Bulletin*, American Water Resource Association, Vol. 19, No. 4.

Madalla G.S., 1983. Limited Dependent and Qualitative Variables in Econometrics, Cambridge University Press, Cambridge.

Maloney, Clarence and Raju, K.V., 1994. Managing Irrigation Together : Practice Policy in India, Sage Publications, New Delhi.

Mas-Collel, A.; Whiston, M. and Green, Jerry, 1995. Micro-economic Theory, Oxford University Press, Oxford.

Mathur P.C. and Saibal S.K., 2002. Participatory Irrigation Management : Integration with System Rehabilitation in Joshi L.K and Hooja, R. (ed.). *Participatory Irrigation Management : Paradigm for 21st Century, Vol. 2,* Rawat Publication, New Delhi.

Meinzen-Dick; Ruth S.; Raju K.V. and Gulati Asok. What Affects Organization and Collective Action for Managing Resources? Evidence from Canal Irrigation Systems in India, *World Development*, Vol. 30, No. 4.

Merry, Douglas, J., 1982. Reorganising Irrigation : Local Level Management in Punjab (Pakistan), Reprint 2, International Irrigation Management Institute, Colombo.

Merry, Douglas, J., 1997. Expanding the Frontiers of Irrigation Management Research, International Irrigation Management Institute, Colombo.

Mitra, Ashok, K. 1997. Economic Aspects of Irrigation Management in Major and Medium Surface Irrigation Systems in India, *Artha Vijnana*, Vol. XXXIX, No. 3.

Moench, Marcus (ed.), 1995. Ground Water Law : The Growing Debate. VIKSAT-Pacific Institute Collaborative Ground Water Project, Ahmedabad.

Nair, S.K.N., 1999. Resource Management for Sustainable Development in Kerala. In Oommen, M.A.(ed.). *Rethinking Development : Kerala's Development Experience*, Delhi : Concept Publishing Company.

Narayana, D. and Nair K.N., 1983. Linking Irrigation with Development : The Kerala Experience. *Economic and Political Weekly*, Vol. 8, Nos. 45-46.

Netto. N., 1990. Public Investments in Irrigation Projects in Kerala : An Analysis of Cost and Organisational Structure. Unpublished M.Phil. Thesis. Centre for Development Studies, Trivandrum.

North, Douglass, C., 1990. Institutions, Institutional Change and Economic Performance, Cambridge University Press, Cambridge.

North, Douglass, C., 1992. Institutions and Economic Theory, *The American Economist*, Vol. 36.

North, Douglass, C. 1993. Institutions and Credible Commitment, *Journal of Institutional and Theoretical Economics*, Vol. 149.

North, Douglas, C. and Thomas, Robert Paul, 1970. An Economic Theory of the Growth of the Western World, *Economic History Review*, Second Series, Vol. 23, No. 1.

Oakerson, Ronald J., 1986. A Model for the Analysis of Common Property Problems. In *Proceedings of the Conference on Common Property Resource Management*, National Research Council, National Academy Press. Washington D.C.

Oakerson, Ronald J., 1992. Analysing the Commons : A Framework in Bromely, Daniel W. (ed.), *Making the Commons Work : Theory , Practice and Policy*, ICS Press, San Francisco.

Olson, Mancur, 1965. The Logic of Collective Action, Public Goods and the Theory of the Groups, Harvard University Press, Cambridge.

Ostrom, Elinor, 1986. An Agenda for the Study of Institutions. *Public Choice*, Vol. 48. No. 1.

Ostrom, Elinor, 1990. Governing the Commons : The Evolution of Institutions for Collective Action, Cambridge University Press, New York.

Ostrom, Elinor, 1992. Crafting Institutions for Self-Governing Irrigation Systems, Institute for Contemporary Studies, California.

Ostrom, Elinor, 1995. Incentives, Rules of the Game and Development, Paper presented at the Annual World Bank Conference on Development Economics. World Bank, Washington, D.C.

Ostrom, Vincent and Ostrom E., 1977. Public Good and Public Choices in Savas, E.S. (ed.). Alternatives for Delivering Public Services : Towards Improved Performance, Westview Press, Colombia.

Ostrom, Elinor and Gardner, Roy, 1993. Coping with Assymetries in the Commons : Self Governing Irrigation Systems can Work. *Journal of Economic Perspectives*, Vol. 7, No. 4.

Palaniswami, 1984. Irrigation Water Management : The Determinants of Canal Water Distribution in India : A Micro Analysis, Agricole Publishing Company, New Delhi.

Panickar, P.G.K., 1980. Recent Trends in the Area under and Production of Rice in Kerala, Working Paper 116. Centre for Development Studies, Trivandrum.

Pant, Niranjan, 1981. Utilisation of Canal Water Below Outlet in Kosi Irrigation Project : Administrative and Community Level Solutions, *Economic and Political Weekly*, Vol. XVI, No. 39.

Parasuraman, S. and Sengupta, S., 2001. World Commission on Dams : Democratic Means for Sustainable Ends, *Economic and Political Weekly*, Vol. XXXVI, No. 21.

Parthasarathy, R., 2000. Participatory Irrigation Management Programme in Gujarat : Institutional and Financial Issues, *Economic and Political Weekly*, Vol. XXXV, Nos. 35-36.

Parthasarathy, R., 2002. Comparing Gujarat with Andhra Pradesh : Reforms in Irrigation Management and Peoples' Participation in Hooja, Rakesh, Pangare, Ganesh and Raju, K.V. (ed.). *Users in Water Management*, Rawat Publications, New Delhi.

Patil, R.K., 1997 : Water Management in Major and Medium Projects, *Artha Vijnana*, Vol. XXXIX, No. 3.

Perry, C.J., 2001. Charging for Irrigation Water : The Issues and Options : With a Case Study from Iran, Research Report 52, International Water Management Institute, Colombo.

Perry, C.J. and Narayanamurthy, 1998. Farmer Responses to Rationed and Uncertain Irrigation Supplies, Research Report 24, International Water Management Institute, Colombo.

Perry, C.J.; Rock, Michael and Seckler, D., 1997. Water as an Economic Good : A Solution, or a Problem?. Research Report 14, International Irrigation Management Institute, Colombo.

Pillai, P.P. (ed.), 1982. Agriculture Development in Kerala, Agricole Publishing Company, New Delhi.

Posner, Richard A., 1977. Economic Analysis of Law, Little Brown, Boston.

Prasad, G.S.G., 2002. Water Markets : Public Resources and Private Appropriation, *Economic and Political Weekly*, Vol. XXXVII, No. 1.

Raju, K.V., 1998. Changing Dynamics of Irrigated Agriculture under Institutional Stress, *The Social Engineer*, Vol. 7, No. 2.

Rao, C.H.H., 2002. Sustainable Use of Water for Irrigation in Indian Agriculture, *Economic and Political Weekly*, Vol. XXXVII, No. 18.

Rao, P.K., 2003. The Economics of Transaction Costs : Theory, Methods and Application, Palgrave MacMillan, New York.

Reddy, S., 1997. Different Sources of Irrigation : A Case Study of Telangana Region, Manak Publications, New Delhi.

Repetto, Robert, 1986. Skimming the Water : Rent-seeking and the Performance of Public Irrigation Systems, Research Report 41. World Resource Institute, Washington DC.

Rinaudo, Jean-Daniel *et al.*, 1997. Linking Water Market Functioning, Access to Water Resources and Farm Production Strategies : Examples from Pakistan, *Irrigation and Drainage Systems*, Vol. 11, No. 3.

Rosegrant, Mark W. and Binswanger, Hans, P., 1994. Markets in Tradable Water Rights : Potential for Efficiency Gains in Developing Country Water Resource Allocation, *World Development*, Vol. 22, No. 11.

Rosegrant M.W., Schleyer, R.G. and Yadav, S.N., 1995. Water Policy for Efficient Agricultural Diversification : Market Based Approaches, *Food Policy*, Vol. 20, No. 3.

Rosegrant, Mark W. and Schleyer, R.G., 1997. Markets in Tradable Water Rights : Lessons for Asia from Chile, Mexico and California in Rosegrant M.W, *et al.* (ed) *Agricultural Sustainability, Growth and Poverty Alleviation in East and South East Asia : Issues and Policies*, IFPRI and ISIS, Washington D.C.

Roth, G., 1987. The Private Provision of Public Services in Developing Countries, Oxford University Press, New York.

Runge, C. Ford, 1986 : Common Property and Collective Action in Economic Development. *World Development*, Vol. 15, No. 5.

Rutherford, Malcom, 1994. Institutions in Economics : The Old and New Institutionalism, Cambridge University Press, Cambridge.

Samad, Madar and Vermillion Douglas, 1999. Assessment of Participatory Management of Irrigation Schemes in Sri Lanka : Partial Reforms, Partial Benefits, Research Report 34, International Water Management Institute, Colombo.

Samad, Madar and Vermillion Douglas, 1999. An Assessment of the Impact of Participatory Management of Irrigation Schemes in Sri Lanka, *International Journal of Water Resources Development*, Vol. 15, No. 1-2.

Sampath, Rajan K., 1992. Issues in Irrigation Pricing in Developing Countries, *World Development*, Vol. 20, No. 7.

Sampath, R.K. and Young R.A., 1990. Social, Economic and Institutional Issues in Third World Irrigation Management, Studies in Water Policy and Management, No.15, Westview Press, Oxford.

Santhakumar, V., 1997. Institutional Lock-in in Natural Resource Management : The Case of Water Resources in Kerala. Working Paper 276, Centre for Development Studies, Trivandrum.

Santhakumar, V.; Rajagoplan, R. and Ambirajan, S., 1995. Planning Kerala's Irrigation Projects : Technological Prejudice and Politics of Hope. *Economic and Political Weekly*, Vol. XXX, No. 12.

Santhakumar, V. and Nair, K.N., 1999. Kerala's Agriculture : Trends and Prospects in Oommen, M.A (ed) *Rethinking Development : Kerala's Development Experience*, Concept Publishing Company, Delhi.

Seabright, S., 1990. Is Co-operation Habit Forming? Mimeo, Cambridge : Cambridge University Press.

Sen, A.K., 1982. Choice, Welfare and Measurement. Basil Blackwell, Oxford.

Sen, A.K., 1987. On Ethics and Economics, Basil Blackwell, Oxford.

Sengupta, Nirmal, 1991. Managing Common Property : Irrigation in India and the Phillipines, Sage, New Delhi.

Sengupta, Nirmal, 2001. A New Institutional Theory of Production : An Application, Sage, New Delhi

Shah, Amita, 2002. "Watershed Programme in Gujarat : A Long Way to Go" in Hirway, Indira, Kashyap, S.P. and Shah, Amita (ed.), *Dynamics of Development in Gujarat*, Concept Publishing Company, New Delhi.

Shah, Tushaar, 1985. Transforming Ground Water Markets into Powerful Instruments of Small Farmer Development : Lessons from the Punjab, Uttar Pradesh and Gujarat. ODI Network Paper 11d, Overseas Development Institute, London.

Shah, Tushaar, 1989. Ground Water Markets : A Review of Issues, Evidence and Policies, Institute of Rural Management, Anand.

Shah, Tushaar, 1993. Ground Water Markets and Irrigation Development : Political Economy and Practical Policy, Oxford University Press, Bombay.

Shah, Tushaar, 2001. Wells and Welfare in the Ganga Basin : Public Policy and Private Initiative in Eastern Uttar Pradesh, India. IWMI Research Report 54, International Water Management Institute, Colombo.

Shah, Tushaar and Ballabh, Vishwa, 1997. Water Markets in North Bihar—Six Village Studies in Muzaffarpur District, *Economic and Political Weekly*, Vol. XXXII, No. 52.

Shah, Tushaar, *et al.*, (2000) : 'Irrigation Management in Pakistan and India : Comparing Notes on Institutions and Policies, Working Paper 4, International Water Management Institute, Colombo.

Shaktivadivel, *et al.*, 2001. Water Scarcity and Managing Seasonal Water Crisis : Lessons from the Kirindi Oya Project in Sri Lanka, Research Report 55, International Water Management Institute, Colombo.

Singh K.K.; Turab-Ul-Hassan; Rao C. Sithapathi and Kemuel George, 1994. Promoting Farmers' Participation : Action Research in Sreeramsagar Project, in Sivamohan, M.V.K., Scott, A., Christopher (ed.), *India : Irrigation Management Partnerships*, Booklinks Corporation, Hyderabad.

Singh, K.K., 2000. Exploration without Commitment : The Story of Participatory Irrigation Management in India, in Joshi, L.K and Hooja, R. (ed.), *Participatory Irrigation Management : Paradigm for 21st Century—Vol. 2*, Rawat Publication, New Delhi.

Srivastava, L.P. and Brewer, Jeffrey D., 1994. Irrigation Management Transfer at Paliganj Canal, Bihar, India, International Irrigation Management Institute, Colombo.

Strosser, P. and Kuper, M., 1994. Water Markets in the Fordvah/Eastern Sadiqia Area : An Answer to Perceived Deficiencies in Canal Water Supply, Working Paper 30, International Irrigation Management Institute, Colombo.

Sugden, Robert, 1986. The Economics of Property Rights, Cooperation and Welfare, Blackwell Ltd., Oxford.

Svendsen, M. and Meinzen-Dick, R., 1997. Irrigation Management Institutions in Transition : A Look Back, A Look Forward. *Irrigation and Drainage Systems*, Vol. 11, No. 2.

Tang, Shui Yan, 1992. Institutions and Collective Action : Self Governance in Irrigation, ICS Press, California.

Tang, Shui Yan, 1993. Integrating Local Participation and Institutional Development : A Transaction Cost Perspective, *Research in Public Administration*, Vol. 2.

Taylor, M., 1987. The Possibility of Co-operation, Cambridge University Press, Cambridge.

Unni, J., 1983. Changes in the Cropping Pattern in Kerala : Some Evidences on Substitution of Coconut for Rice, 1960-61 to 1978-79. *Economic and Political Weekly*, Vol. XXVIII, No. 39.

Uphoff, Norman, 1992. Local Institutions and Participation for Sustainable Development, Gatekeeper Series 31, IIED, London.

Vermillon, Douglas L., 1991. The Turnover and Self Management of Irrigation Institutions in Developing Countries, A Discussion Paper, International Irrigation Management Institute, Colombo.

Vermillon, Douglas L.; Samad, Madar, Pusposutardjo, S.; Arif, Srijit, S. and Rochdyanato, S., 2000. An Assessment of the Small-Scale Irrigation Management Turnover Programme in Indonesia, Research Report 38, International Irrigation Management Institute, Colombo.

Viswanathan, P.K., 1997. Irrigation Projects in Kerala : A review of past performance and future perspectives. Paper presented in *Dr. T.N. Krishnan Memorial Seminar on Development Experience of South Indian States in a Comparative Setting,* Centre for Development Studies, September 7-9, Trivandrum.

Wade, Robert, 1975. Administration and the Distribution of Irrigation Benefits. *Economic and Political Weekly,* Vol. XXI, Nos. 44-45.

Wade, Robert, 1978. Water Supply as an Instrument of Agricultural Policy : A Case Study. *Economic and Political Weekly,* Vol. XIII, No. 12.

Wade, Robert, 1980. India's Changing Strategy of Irrigation Development, in Coward Jr. and Walter E. (ed) Irrigation and Agricultural Development in Asia : Perspectives from the Social Sciences, Cornell University Press, London.

Wade, Robert, 1982. The System of Administrative and Political Corruption : Canal Irrigation in South India. *Journal of Development Studies,* Vol. 18, No. 3.

Wade, Robert, 1985. The Market for Public Office : Why the Indian State is not Better at Development. *World Development,* Vol. 13, No. 4.

Wade, Robert, 1987. The Operation of an Indian Irrigation Bureaucracy. Agricultural and Rural Development Department. Mimeo, The World Bank, Washington D.C.

Wade, Robert, 1988a. The Management of Common Property Resources : Finding a Co-operative Solution, World Bank, Washington DC.

Wade, Robert, 1988b. The Management of Irrigation Systems : How to Evoke Trust and Avoid Prisoners' Dilemma, *World Development,* Vol. 16, No. 4.

Wade, Robert and Chambers, Robert, 1980. Managing the Main System : Canal Irrigation's Blind Spot. *Economic and Political Weekly,* Vol. XV, No. 39.

Wickham, T.H. and Valera, 1979. Practices and Accountability for Better Water Management in Taylor, D.C. and Wickham, T.H. (ed.). *Irrigation Policy and Management of Irrigation Systems in South East Asia,* Agricultural Development Council, Bangkok.

Williamson, O.E., 1979. Transaction Cost Economics : The Governance of Contractual Relations, *Journal of Law Economics*, Vol. 22, No. 2.

Williamson, O.E., 1985. The Economic Institutions of Capitalism, Collier Macmillan Publishers, London.

World Bank, 1994. World Development Report 1994 : Infrastructure Development, Oxford University Press, New York.

Index